Reviews of Environmental Contamination and Toxicology

VOLUME 161

Springer
New York
Berlin
Heidelberg
Barcelona
Hong Kong
London
Milan
Paris
Singapore
Tokyo

Reviews of Environmental Contamination and Toxicology

Continuation of Residue Reviews

Editor
George W. Ware

VOLUME 161

Springer

Springer-Verlag
New York: 175 Fifth Avenue, New York, NY 10010, USA
Heidelberg: Postfach 10 52 80, 69042 Heidelberg, Germany

Printed in the United States of America.
ISBN 978-1-4419-3152-8
ISSN 0179-5953

Printed on acid-free paper.

Foreword

International concern in scientific, industrial, and governmental communities over traces of xenobiotics in foods and in both abiotic and biotic environments has justified the present triumvirate of specialized publications in this field: comprehensive reviews, rapidly published research papers and progress reports, and archival documentations. These three international publications are integrated and scheduled to provide the coherency essential for nonduplicative and current progress in a field as dynamic and complex as environmental contamination and toxicology. This series is reserved exclusively for the diversified literature on "toxic" chemicals in our food, our feeds, our homes, recreational and working surroundings, our domestic animals, our wildlife and ourselves. Tremendous efforts worldwide have been mobilized to evaluate the nature, presence, magnitude, fate, and toxicology of the chemicals loosed upon the earth. Among the sequelae of this broad new emphasis is an undeniable need for an articulated set of authoritative publications, where one can find the latest important world literature produced by these emerging areas of science together with documentation of pertinent ancillary legislation.

Research directors and legislative or administrative advisers do not have the time to scan the escalating number of technical publications that may contain articles important to current responsibility. Rather, these individuals need the background provided by detailed reviews and the assurance that the latest information is made available to them, all with minimal literature searching. Similarly, the scientist assigned or attracted to a new problem is required to glean all literature pertinent to the task, to publish new developments or important new experimental details quickly, to inform others of findings that might alter their own efforts, and eventually to publish all his/her supporting data and conclusions for archival purposes.

In the fields of environmental contamination and toxicology, the sum of these concerns and responsibilities is decisively addressed by the uniform, encompassing, and timely publication format of the Springer-Verlag (Heidelberg and New York) triumvirate:

Reviews of Environmental Contamination and Toxicology [Vol. 1 through 97 (1962–1986) as Residue Reviews] for detailed review articles concerned with any aspects of chemical contaminants, including pesticides, in the total environment with toxicological considerations and consequences.

Bulletin of Environmental Contamination and Toxicology (Vol. 1 in 1966) for rapid publication of short reports of significant advances and discoveries in the fields of air, soil, water, and food contamination and pollution as well as methodology and other disciplines concerned with the introduction, presence, and effects of toxicants in the total environment.

Archives of Environmental Contamination and Toxicology (Vol.1 in 1973) for important complete articles emphasizing and describing original experimental or theoretical research work pertaining to the scientific aspects of chemical contaminants in the environment.

Manuscripts for *Reviews* and the *Archives* are in identical formats and are peer reviewed by scientists in the field for adequacy and value; manuscripts for the *Bulletin* are also reviewed, but are published by photo-offset from camera-ready copy to provide the latest results with minimum delay. The individual editors of these three publications comprise the joint Coordinating Board of Editors with referral within the Board of manuscripts submitted to one publication but deemed by major emphasis or length more suitable for one of the others.

Coordinating Board of Editors

Preface

Thanks to our news media, today's lay person may be familiar with such environmental topics as ozone depletion, global warming, greenhouse effect, nuclear and toxic waste disposal, massive marine oil spills, acid rain resulting from atmospheric SO_2 and NO_x, contamination of the marine commons, deforestation, radioactive leaks from nuclear power generators, free chlorine and CFC (chlorofluorocarbon) effects on the ozone layer, mad cow disease, pesticide residues in foods, green chemistry or green technology, volatile organic compounds (VOCs), hormone- or endocrine-disrupting chemicals, declining sperm counts, and immune system suppression by pesticides, just to cite a few. Some of the more current, and perhaps less familiar, additions include *xenobiotic transport, solute transport, Tiers 1 and 2, USEPA to cabinet status, and zero-discharge.* These are only the most prevalent topics of national interest. In more localized settings, residents are faced with leaking underground fuel tanks, movement of nitrates and industrial solvents into groundwater, air pollution and "stay-indoors" alerts in our major cities, radon seepage into homes, poor indoor air quality, chemical spills from overturned railroad tank cars, suspected health effects from living near high-voltage transmission lines, and food contamination by "flesh-eating" bacteria and other fungal or bacterial toxins.

It should then come as no surprise that the '90s generation is the first of mankind to have become afflicted with *chemophobia*, the pervasive and acute fear of chemicals.

There is abundant evidence, however, that virtually all organic chemicals are degraded or dissipated in our not-so-fragile environment, despite efforts by environmental ethicists and the media to persuade us otherwise. However, for most scientists involved in environmental contaminant reduction, there is indeed room for improvement in all spheres.

Environmentalism is the newest global political force, resulting in the emergence of multi-national consortia to control pollution and the evolution of the environmental ethic. Will the new politics of the 21st century be a consortium of technologists and environmentalists or a progressive confrontation? These matters are of genuine concern to governmental agencies and legislative bodies around the world, for many serious chemical incidents have resulted from accidents and improper use.

For those who make the decisions about how our planet is managed, there is an ongoing need for continual surveillance and intelligent controls to avoid endangering the environment, the public health, and wildlife. Ensuring safety-in-use of the many chemicals involved in our highly industrialized culture is a dynamic challenge, for the old, established materials are continually being displaced by newly developed molecules more acceptable to federal and state regulatory agencies, public health officials, and environmentalists.

Adequate safety-in-use evaluations of all chemicals persistent in our air, foodstuffs, and drinking water are not simple matters, and they incorporate the judgments of many individuals highly trained in a variety of complex biological, chemical, food technological, medical, pharmacological, and toxicological disciplines.

Reviews of Environmental Contamination and Toxicology continues to serve as an integrating factor both in focusing attention on those matters requiring further study and in collating for variously trained readers current knowledge in specific important areas involved with chemical contaminants in the total environment. Previous volumes of *Reviews* illustrate these objectives.

Because manuscripts are published in the order in which they are received in final form, it may seem that some important aspects of analytical chemistry, bioaccumulation, biochemistry, human and animal medicine, legislation, pharmacology, physiology, regulation, and toxicology have been neglected at times. However, these apparent omissions are recognized, and pertinent manuscripts are in preparation. The field is so very large and the interests in it are so varied that the Editor and the Editorial Board earnestly solicit authors and suggestions of underrepresented topics to make this international book series yet more useful and worthwhile.

Reviews of Environmental Contamination and Toxicology attempts to provide concise, critical reviews of timely advances, philosophy, and significant areas of accomplished or needed endeavor in the total field of xenobiotics in any segment of the environment, as well as toxicological implications. These reviews can be either general or specific, but properly they may lie in the domains of analytical chemistry and its methodology, biochemistry, human and animal medicine, legislation, pharmacology, physiology, regulation, and toxicology. Certain affairs in food technology concerned specifically with pesticide and other food-additive problems are also appropriate subjects.

Justification for the preparation of any review for this book series is that it deals with some aspect of the many real problems arising from the presence of any foreign chemical in our surroundings. Thus, manuscripts may encompass case studies from any country. Added plant or animal pest-control chemicals or their metabolites that may persist into food and animal feeds are within this scope. Food additives (substances deliberately added to foods for flavor, odor, appearance, and preservation, as well as those inadvertently added during manufacture, packing, distribution, and storage) are also considered suitable review material. Additionally, chemical contamination in any manner of air, water, soil, or plant or animal life is within these objectives and their purview.

Normally, manuscripts are contributed by invitation, but suggested topics are welcome. Preliminary communication with the Editor is recommended before volunteered review manuscripts are submitted.

Department of Entomology
University of Arizona
Tucson, Arizona

G.W.W.

Table of Contents

Rev Environ Contam Toxicol 161:1–156

Nitroaromatic Munition Compounds: Environmental Effects and Screening Values

Sylvia S. Talmage, Dennis M. Opresko, Christopher J. Maxwell, Christopher J.E. Welsh, F. Michael Cretella, Patricia H. Reno, and F. Bernard Daniel

Contents

Communicated by George W. Ware
S.S. Talmage (✉)·D.M. Opresko·C.J. Maxwell·C.J.E. Welsh·F.M. Cretella·P.H. Reno
Life Sciences Division, Oak Ridge National Laboratory, 1060 Commerce Park, Oak Ridge, TN 37830, U.S.A.
F.B. Daniel U.S. Environmental Protection Agency, Ecological Exposure Research Division, National Exposure Research Laboratory, 26 W. Martin Luther King Drive, Cincinnati, OH 45268, U.S.A.

I. Introduction

Nitroaromatic compounds, including 2,4,6-trinitrotoluene (TNT), hexahydro-1,3,5-trinitro-1,3,5-triazine (RDX), octahydro-1,3,5,7-tetranitro-1,3,5-tetrazocine (HMX), N-methyl-N,2,4,6-tetranitroaniline (tetryl), and associated byproducts and degradation products, were released to the environment during manufacturing and load, assembly, and pack (LAP) processes at U.S. Army Ammunition Plants (AAPs) and other military facilities. As a result of the release of these nitroaromatic compounds into the environment, many AAPs have been placed on the National Priorities List for Superfund cleanup (Fed. Reg. 60:20330). Many of these sites cover a wide expanse of relatively undisturbed land and provide diverse habitats that support a variety of aquatic and terrestrial species. Nitroaromatics are potentially toxic to the indigenous species at these sites and present a significant concern for site remediation. Table 1 presents an overview of ranges of detected concentrations of the nitroaromatic compounds in groundwater, surface water, sediment, and soil at military and manufacturing sites. In addition to sites described in the text, the U.S. Army Installation Restoration

Table 1. Concentration ranges of nitroaromatic munitions compounds at military installations.

Chemical	Groundwater (μg/L)	Surface water (μg/L)	Sediment (mg/kg)	Soil (mg/kg)
2,4,6-Trinitrotoluene	0.4–21,960	1.0–3375	6.7–711,000	0.08–87,000[a]
1,3,5-Trinitrobenzene	0.2–7720[b]	3.0–97	<0.1–5.1	0.08–67,000
1,3-Dinitrobenzene	0.74–704	0.6–6.6	0.59	0.06–45.2[c]
3,5-Dinitroaniline	ND	ND	ND	0.08–0.67[d]
2-Amino-4,6-dinitrotoluene	ND	ND	0.07[e]	<0.1–37
RDX	0.5–36,000[f]	2.6–224[e]	<1–43,000	0.7–74,000[g]
HMX	1.3–4200[h]	1.9–67[i]	≤1.27	0.7–5700[c]
Tetryl	0.6–235	≤132	0.25–1.3	0.5–84,400[j]

ND, no data.

Source: U.S. Army Installation Restoration Data Management Information System, Aberdeen Proving Ground, MD, except as otherwise noted.

[a]Simini et al. (1995).
[b]US Army (1987b).
[c]Walsh and Jenkins (1992).
[d]Grant et al. (1995).
[e]Spanggord et al. (1981).
[f]Tucker et al. (1985).
[g]Newell (1984).
[h]Gregory and Elliott (1987).
[i]Stidham (1979).
[j]Small and Rosenblatt (1974).

Data Management Information System lists soil concentrations at many sites. These values are also listed in Table 1.

At present, no scheme exists for evaluating the potential hazards to aquatic organisms and terrestrial wildlife from munitions chemicals commonly found in water and soil at these sites. To determine if concentrations at a site might be harmful to the indigenous species, the maximum measured media-specific concentration can be compared to a criterion or screening benchmark. The criterion or benchmark is a concentration that should not result in adverse ecological effects to the populations of indigenous species. For some media, e.g., surface water and sediment, U.S. Environmental Protection Agency (USEPA) methods are available for developing water quality criteria or screening benchmarks. For mammalian species, lowest-observed-adverse-effect levels (LOAELs) or no-observed-adverse-effect levels (NOAELs) for food and water can be calculated in the same manner that these levels are set for humans. For terrestrial plants and soil invertebrates and microbes, lowest- or no-observed-effect concentrations (LOECs or NOECs) can be determined from the available data.

The USEPA refers to media-specific benchmark values to be used for screening purposes at Superfund sites as Ecotox thresholds (USEPA 1996). Ecotox thresholds are defined as contaminant concentrations above which there is sufficient concern regarding adverse ecological effects to warrant further site investigations. For surface water and sediment, the preferred respective ecotox thresholds are the chronic ambient Water Quality Criteria (WQC) and the Sediment Quality Criteria (SQC) based on equilibrium partitioning (EqP). The criteria and screening benchmarks presented here can be considered ecotox thresholds. Suter and Tsao (1996) suggest that acute and chronic ambient WQC can be used as upper and lower screening benchmarks, respectively. Exceedance of an upper screening benchmark indicates that a contaminant is of concern; exceedance of a lower screening benchmark indicates the need for further assessment.

This review presents a summary and analysis of available data on eight nitroaromatic compounds including environmental concentrations, environmental fate and transport processes, and ecotoxicity and bioaccumulation for aquatic and terrestrial biota. For those groups of organisms for which there are sufficient data, ecological criteria and screening benchmarks were developed. These criteria and screening benchmarks were developed by staff at Oak Ridge National Laboratory (ORNL) under a project jointly sponsored by the U.S. Army and the USEPA. The methodologies for development of the screening criteria and benchmarks are discussed in the following pages.

A. Calculation of Aquatic Criteria/Screening Benchmarks

Numerical national ambient WQC, developed by the USEPA Office of Water, may be used as screening benchmarks for freshwater organisms. Although WQC have not been calculated by USEPA for the munitions compounds discussed here, the methods established by USEPA can be applied to the available data on these compounds.

1. Tier I Water Quality Criteria. Calculation of Tier I WQC for protection of aquatic organisms is based on USEPA guidelines (Stephan et al. 1985). Criteria are of two types, acute and chronic. The acute WQC or Criterion Maximum Concentration (CMC) is the highest 1-hr average concentration that should not result in unacceptable effects on aquatic organisms and their uses. The chronic WQC or Criterion Continuous Concentration (CCC) is the highest 4-d average concentration that should not cause unacceptable toxicity during a long-term exposure. These thresholds should not be exceeded more than once every 3 yr. Development of acute and chronic WQC requires results of eight acute toxicity tests and three chronic toxicity tests. For the acute WQC, acceptable acute toxicity test results must be available for eight different aquatic families: (1) Salmonidae, (2) a second family in the class Osteichthyes, (3) a third family in the phylum Chordata, (4) a planktonic crustacean, (5) a benthic crustacean, (6) an insect, (7) a family in a phylum other than Arthropoda, and (8) a family in any order of insect or any phylum not already represented.

If these eight data requirements are met, a Final Acute Value (FAV) is calculated. Several steps are involved in calculating the FAV. First, for each species for which more than one acute value is available, the Species Mean Acute Value (SMAV) should be calculated as the geometric mean of the values. For each genus for which one or more SMAVs are available, the Genus Mean Acute Value (GMAV) should be calculated as the geometric mean of the SMAVs. The GMAVs for each family are ordered from high to low with ranks, R, from 1 for the lowest to N for the highest assigned. The cumulative probability, P, for each GMAV is calculated as $R/(N + 1)$. The four GMAVs that have cumulative probabilities closest to 0.05 are then selected. If there are fewer than 59 GMAV, these will be the 4 lowest. Using the GMAV and P values, the FAV is calculated as follows:

$$\text{FAV} = e^{A}$$

where

$$A = S\left(\sqrt{0.05}\right) + L$$

$$L = \frac{\left[\Sigma\ (\ln \text{GMAV}) - S\left(\Sigma\ \left(\sqrt{P}\right)\right)\right]}{4}$$

and

$$S^2 = \frac{\Sigma[(\ln \text{GMAV})^2] - \left[\frac{(\Sigma \ln \text{GMAV})^2}{4}\right]}{\Sigma(P) - \frac{[(\Sigma\sqrt{P})^2]}{4}}$$

The CMC, which is one-half of the FAV, is the fifth percentile of the distribution of 48- to 96-hr LC_{50} values or equivalent effective concentration (EC_{50}) values for each criterion chemical. The CMC is intended to correspond to concentrations that would cause less than 50% mortality in 5% of exposed populations in a brief exposure.

Following calculation of the FAV, a Final Chronic Value (FCV) can be calculated in the same manner as the FAV or, if chronic tests are not available for eight families, the FCV can be calculated by dividing the FAV by the Final Acute/Chronic Ratio (FACR). In the latter case, for each Chronic Value (CV) for which at least one corresponding appropriate acute value is available, an acute/chronic ratio is calculated. The CV is the geometric mean of the highest concentration that did not cause an unacceptable effect (NOEC) and the lowest concentration that did cause an unacceptable adverse effect (LOEC) in a chronic study. A minimum of three CVs are needed to derive the ratio. Tests should have been conducted under flow-through conditions (static tests may be used for daphnids), and concentrations should be measured rather than nominal. Chronic tests may be life cycle, partial life cycle, or early life stage. Acute and chronic tests should have been conducted in the same laboratory, but tests conducted in different laboratories may be used. For each species, the mean acute/chronic ratio is calculated as the geometric mean of all acute/chronic ratios for that species. If no major trend is apparent in the acute/chronic ratios and they are within a factor of 10, the FACR is calculated as the geometric mean of all species mean acute/chronic ratios.

The Final Plant Value (FPV) is defined as the lowest result from a test with an important aquatic plant species in which the concentration of test material was measured and the endpoint was biologically important. The test should be of 96-hr duration with a species of algae or a chronic test with an aquatic vascular plant. Although test procedures with plants are not well developed, results of toxicity tests with plants usually indicate that criteria which protect aquatic animals and their uses will protect aquatic plants and their uses.

The chronic WQC or CCC is the lowest of three values: the FCV, the FPV, and the final residue value. The final residue value is defined as the lowest of the residue values that are obtained by dividing maximum permissible tissue concentrations, defined as either (a) a U.S. Food and Drug Administration action level or (b) the maximum acceptable dietary intake derived from a chronic wildlife study, by appropriate bioconcentration or bioaccumulation factors. No final residue values are available for the munitions compounds. The few experimental bioconcentration studies for nitroaromatic munitions compounds suggest that these compounds do not bioconcentrate. Action levels were not considered here because they pertain to human consumption of aquatic organisms, and all screening benchmarks are based on ecotoxicity. The FAV and the FPV are presented here as standalone values. Screening benchmarks for food or water intake for wildlife species were calculated separately. Therefore, the FCV as calculated here may be considered the same as the chronic WQC or CCC, which is the preferred USEPA ecotox threshold for surface water.

2. Tier II Water Quality Criteria. If Tier I WQC have not been derived by USEPA or if toxicity data are insufficient to derive Tier I acute and chronic WQC for protection of aquatic organisms according to the methods of Stephan et al. (1985), the USEPA *Proposed Water Quality Guidance for the Great Lakes System* (USEPA 1993a) presents a method for derivation of benchmarks (analogous to the CMC and CCC) with fewer data points than the number required for Tier I. These guidance values are referred to as secondary or Tier II values.

Tier II methodology uses adjustment factors termed Secondary Acute Factors (SAFs) to calculate Tier II values. The Secondary Acute Value (SAV), analogous to the Tier I FAV, is calculated by dividing the lowest genus mean acute value (GMAV) in the database by the SAF (the database must contain a GMAV for a species of daphnid). USEPA (1993a) lists SAFs for use in Tier II calculations; these SAFs are based on the number of satisfied data requirements for Tier I calculations (Table 2). The Secondary Maximum Concentration (SMC), analogous to the acute WQC, is one-half of the SAV.

Following calculation of the SAV, a Secondary Chronic Value (SCV), analogous to the Tier I FCV, can be calculated depending on the availability of data. If data on eight species are not available, acute/chronic ratios (ACR) for the species of interest can be calculated. The geometric mean of the ACRs is designated the secondary acute/chronic ratio (SACR). The SCV is the SAV divided by the SACR. The default ACR is 18.

Table 2. USEPA secondary acute factors for Tier II acute aquatic criteria.

Number of satisfied minimum data requirements	1	2	3	4	5	6	7
Secondary acute factor	20	13	8.6	6.5	5.0	4.0	3.6

Source: USEPA (1993a).

3. Lowest Chronic Values. For some of the chemicals listed in the USEPA WQC summary (USEPA 1986), data were insufficient to calculate Tier I or Tier II values. For those chemicals for which there are not enough data to calculate a criterion but for which at least one chronic value is available, the LOEC for a species of daphnid or fish is listed. For comparison purposes, lowest chronic values for the munitions compounds are listed along with screening benchmarks in this review.

4. Sediment Quality Benchmarks. There are many different approaches to the development of safe values for contaminants in sediments, of which three—the measurement of interstitial (pore) water, spiked sediment toxicity tests, and equilibrium partitioning (EqP)—have been recommended (Jones et al. 1996). For the nonionic organic chemicals presented here, the EqP approach (Di Toro et al. 1991; USEPA 1993b) was used to develop sediment benchmarks. These values are called sediment quality criteria (SQC), but because they have not been approved by the USEPA Science Advisory Board and they incorporate Tier II values as just described, they are referred to here as sediment quality benchmarks (SQB).

The basic assumptions, methodology, and uncertainties in the EqP method are outlined in Di Toro et al. (1991) and USEPA (1993b). The basic calculation uses the chronic FCV (a concentration deemed safe for aquatic organisms) together with correction factors for the effects of organic carbon, the predominant phase for sorption of hydrophobic chemicals. The procedure is as follows. The SQB is computed using the WQC FCV (or SCV if a FCV cannot be calculated) (in mg/L) and the partition coefficient K_p (in L/kg sediment) between water and sediment.

$$SQB = K_p \times FCV$$

The partition coefficient is the ratio of the sediment concentration, C_s, to pore water concentration, C_d. The sorption capacity of the sediment is determined by the mass fraction of organic carbon in the sediment (f_{oc}) and is given by

$$K_p = \frac{C_s}{C_d} = f_{oc} \times K_{oc}$$

where K_{oc} is the partition coefficient for sediment organic carbon. The relationship applies to sediments with $f_{oc} > 0.2\%$ by weight. If the K_{oc} is not available, USEPA (1993b) recommends using a reliably measured or calculated octanol/water partition coefficient, K_{ow}, to estimate the K_{oc}. The K_{oc} may be estimated from the K_{ow} for the chemical using the following equation (Di Toro 1985).

$$\log_{10}(K_{oc}) = 0.00028 + 0.983\ \log_{10}(K_{ow})$$

Therefore, the SQB becomes

$$SQB = f_{oc} \times K_{oc} \times FCV$$

This equation is linear in the organic carbon fraction, f_{oc}, and the relationship can be expressed as

$$\frac{SQB}{f_{oc}} = K_{oc} \times FCV$$

If the organic carbon-normalized SQB, SQB_{oc}, is defined as SQB/f_{oc}, then

$$SQB_{oc} = K_{oc} \times FCV$$

Normalization to a sediment organic carbon content (SQB_{oc}) as recommended by USEPA is more useful than a SQB based on dry weight in that it allows comparison among sites when the f_{oc} in sediment at a particular site is known. To compare the SQB_{oc} to dry weight concentrations at a particular site, the dry weight concentration and the organic carbon concentration at the site must be known; the conversion is mg chemical/kg_{oc} = (mg chemical $_{\text{dry weight}} \times 100)/f_{oc}$. The organic carbon-normalized concentration can then be compared with the SQB_{oc}.

B. Calculation of Terrestrial Screening Benchmarks

1. Mammals. The general method used by ORNL in estimating screening benchmarks for wildlife is based on USEPA methodology for deriving human toxicity values (e.g., reference values, reportable quantities, and unit risks for carcinogenicity) from laboratory animal data (Sample et al. 1996). In the same manner that safe doses of contaminants for humans are based on studies using laboratory animals, reference doses or screening benchmarks for wildlife may be calculated by extrapolation among mammalian species. Screening benchmarks are safe exposure levels (NOAELs) for environmental media (food and water intake) for wildlife. In this approach a NOAEL for population-related effects (e.g., survival, reproductive function) is identified from a study conducted with a species of wildlife or laboratory animal, and the equivalent NOAEL for other species of wildlife (wildlife NOAEL) is obtained by scaling the test data (test NOAEL) on the basis of differences in body size according to the following equation (USEPA 1995a):

$$\text{Wildlife NOAEL} = \text{test NOAEL} \times \left[\frac{\text{test bw}}{\text{wildlife bw}}\right]^{1/4}$$

where: wildlife bw = body weight of wildlife species in kg
test bw = body weight of test species in kg
test NOAEL = experimental dose in mg/kg/d

The methodology for derivation of NOAELs is as follows. In cases in which only a LOAEL is available, the NOAEL is estimated as being equivalent to 1/10th of the LOAEL. If the only available data are a NOAEL (or a LOAEL) for a subchronic exposure (approximately 3 mon to 1 yr), then the equivalent NOAEL or LOAEL for a chronic exposure is estimated as being 1/10th of the value for the subchronic exposure. The screening benchmarks for wildlife derived here are conservative because they are based on NOAELS. For a baseline risk assessment, LOAELs may be more appropriate and may be derived in the same manner.

The dietary screening benchmark (*Cf*, the chemical concentration in food in mg/kg) that would result in a dose equivalent to the NOAEL (assuming no other exposure through other environmental media) is calculated from the food factor *f*, the amount of food consumed/unit bw/d:

$$\mathrm{Cf} = \frac{\text{wildlife NOAEL}}{f}$$

Food factors for the species of wildlife used in the calculations are shown in Table 3; data sources are given in Sample et al. (1996).

The drinking water screening benchmark (Cw, concentration of the chemical in mg/L) that would result in a dose equivalent to the NOAEL (assuming no other exposure through other environmental media) is calculated from the water factor ω, the amount of water consumed/unit bw/d: water factors for the wildlife used in the calculations are listed in Table 3.

$$\mathrm{Cw} = \frac{\text{wildlife NOAEL}}{\omega}$$

If a species (such as mink) feeds primarily on aquatic organisms, and the concentration of the contaminant in the food is proportional to the concentration in the water, then the wildlife NOAEL and body weight, its food intake rate (F, in kg/d) and water intake rate (W, in L/d), and the aquatic life bioaccumulation

Table 3. Reference values for food and water intake for selected wildlife species.

Species	Body weight (kg)	Food intake (kg/d)	Food factor, *f*	Water intake (L/d)	Water factor, ω
Shorttail shrew	0.015	0.009	0.6	0.0033	0.22
White-footed mouse	0.022	0.0034	0.155	0.0066	0.3
Meadow vole	0.044	0.005	0.114	0.006	0.136
Cottontail rabbit	1.2	0.237	0.198	0.116	0.097
Mink	1.0	0.137	0.137	0.099	0.099
Red fox	4.5	0.45	0.1	0.38	0.084
Whitetail deer	56.5	1.74	0.031	3.7	0.065

Source: Sample et al. (1996).

factor (BAF) can be used to derive an overall Cw that incorporates both water and food consumption:

$$Cw = \frac{NOAEL_W \times bw_W}{W + (F \times BAF)}$$

The BAF is the ratio of the concentration of the contaminant in tissue (mg/kg) to its concentration in water (mg/L), where the organism and the prey are exposed. The BAF can be predicted by multiplying the bioconcentration factor (BCF) by the appropriate food chain multiplying factor (FCM).

$$BAF = BCF \times FCM$$

The FCM is a function of the log K_{ow} of the contaminant and the prey trophic level. Because mink feed on small fish, the FCM for trophic level 3 is used in the calculation. All the munitions chemicals discussed in this review have $K_{ow} \leq 3.9$. The FCM for all trophic levels for chemicals with $K_{ow} \leq 3.9$ is 1.0 (USEPA 1993a).

The BCF can be estimated from the log K_{ow} for the chemical by the equation given in Lyman et al. (1982):

$$\log BCF = 0.76 \log K_{ow} - 0.23$$

For chemicals with sufficient data, screening benchmarks were calculated for seven wildlife species. These species—shorttail shrew, white-footed mouse, meadow vole, cottontail rabbit, mink, red fox, and whitetail deer—represent species found throughout the United States. Body weights for selected wildlife species are given in Table 3; data sources are listed in Sample et al. (1996).

2. Plants. In the absence of criteria for terrestrial plants, LOEC values from the literature can be used to screen chemicals of potential concern for phytotoxicity (Will and Suter 1995a). These LOECs described herein are based on growth and yield parameters, in particular, a 20% reduction in growth or yield was used as a level of significant effects for the screening benchmarks. Thus, the LOEC is the lowest applied concentration of a chemical that gave approximately 20% reduction in the measured response. In some cases, only NOECs or thresholds are reported in the reviewed studies. Benchmarks are based on responses in both soil and soil solutions.

3. Soil Invertebrates. According to the method of Will and Suter (1995b), a 20% reduction in growth, reproduction, or activity can be used as the threshold for significant effects on soil invertebrates. Where the data were available, screening benchmarks were established for both earthworms and other soil invertebrates.

4. Soil Heterotrophic Processes. According to the method of Will and Suter (1995b), a 20% reduction in microbial activity can be used as a screening benchmark for microbial heterotrophs. Microbial processes include growth, respiration, and enzyme activities.

C. Data Deficiencies

1. Aquatic Data. For some of the compounds discussed here, only a few tests were available. Data deficiencies result from (1) the chemical nature of the nitroaromatic munitions compounds and (2) lack of toxicity testing. The nitroaromatic munitions compounds are of low solubility and volatility and are stable in the absence of light and moisture. However, TNT, RDX, HMX, and tetryl are subject to photolysis during testing, especially in the presence of light necessary for some tests. Because TNT was not stable under static testing conditions and the products of photolysis were less toxic than the parent compound, results of static toxicity tests with TNT may be underestimates. Some of the compounds are also subject to hydrolysis, although at a slower rate than that for photolysis. Aquatic testing data were available for all chemicals except tetryl.

With one exception, the screening benchmarks derived in this review were calculated according to Tier II guidelines, indicating the lack or inadequacy of studies needed for Tier I guidelines. Acute toxicity tests with eight species were found only for TNT. In most cases testing procedures followed those outlined in the USEPA guidelines. However, tests performed before guideline publication may deviate from the guideline procedures, e.g., the use of static tests instead of the prescribed flow-through tests. In a few cases, test results were either difficult to interpret due to inconsistent results or inappropriate for derivation of benchmarks. In spite of these data constrictions, the best, and in some cases only, available data were used to calculate the screening benchmarks. The data used and exceptions to the testing protocol are noted in each section on calculations. Because some of the data do not meet USEPA data guidelines, these screening benchmarks should be considered as best estimates based on the available data.

The EqP approach was utilized to calculate screening benchmarks for sediment-associated organisms. This method is not applicable to compounds such as 3,5-dinitroaniline (DNA) and 2-amino-4,6-dinitrotoluene (2-ADNT) that may ionize under environmental conditions. An alternate approach must be used for these compounds. The data deficiencies in the aquatic tests are applicable to the sediment benchmarks as the sediment benchmarks are based on the FCV or SCV.

2. Terrestrial Data. Oral toxicity testing data utilizing mammalian laboratory species were located for all chemicals except DNA and 2-ADNT. In most cases multiple subchronic or chronic studies with consistent endpoints and toxicity values were located. No subchronic or chronic feeding studies were found for avian species.

Data on terrestrial plant LOECs were found for four of the eight chemicals:

TNT, 2-ADNT, RDX, and tetryl. Data on the toxicity of the munitions compounds to soil invertebrates and soil heterotrophic processes were sparse.

II. 2,4,6-Trinitrotoluene

2,4,6-Trinitrotoluene (TNT) is a military-unique compound produced at munition production sites. TNT was produced and used extensively in explosives during World War I and II. It has found wide application as a high explosive in shells, bombs, grenades, demolition explosives, and propellant compositions (U.S. Army 1967). TNT is relatively insensitive to shock and must be exploded by a detonator. Small amounts may be used in industrial explosive applications and as a chemical intermediate in the manufacture of dyestuffs and photographic chemicals (ATSDR 1995a). General chemical and physical properties are presented in Table 4.

TNT is manufactured by a continuous process in which nitric acid and oleum are used to nitrate toluene in six to eight stages. The crude TNT is purified with

Table 4. Chemical and physical properties of 2,4,6-trinitrotoluene (TNT).

Synonyms	sym-Trinitrotoluene alpha-Trinitrotoluene 2-Methyl-1,3,5-trinitrobenzene 1-Methyl-2,4,6-trinitrobenzene	HSDB 1995a; Budavari et al. 1996
CAS number	118-96-7	HSDB 1995a
Molecular weight	227.13	Budavari et al. 1996
Physical state	Colorless to pale yellow monoclinic crystals	Budavari et al. 1996
Chemical formula	$C_7H_5N_3O_6$	Budavari et al. 1996
Structure	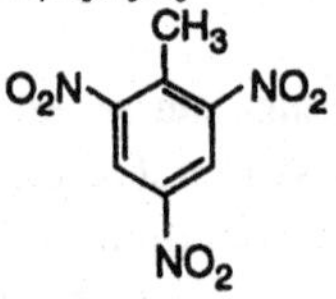	Budavari et al. 1996
Water solubility (20 °C)	130 g/L	Ryon 1987; HSDB 1995a
Specific gravity (20 °C)	1.65	Budavari et al. 1996
Melting point	80.1 °C	Budavari et al. 1996
Boiling point	240 °C	Budavari et al. 1996
Vapor pressure (20 °C)	1.99×10^{-4} mm Hg	HSDB 1995a
Partition coefficients		
Log K_{ow}	1.6, 2.2, 2.7	ATSDR 1995a
	1.84	Ryon 1987
	2.03	Liu et al. 1983a
Log K_{oc}	3.2	Spanggord et al. 1980b
Henry's law constant (20 °C)	4.57×10^{-7} atm-m^3/mole	HSDB 1995a

aqueous sodium sulfite (sellite). Impurities, representing approximately 3% of the final product, include the unsymmetrical isomers of TNT, dinitrotoluene isomers, and oxidation products (Small and Rosenblatt 1974; Ryon et al. 1984).

A. Environmental Fate

1. Sources and Occurrences. TNT is released to the environment from munitions production and processing facilities. It has been identified at 19 National Priorities List sites across the U.S. (ATSDR 1995a); these include AAPs as well as LAP plants. During World War II, 5 AAPs and 25 LAP plants were active. TNT is also present in the environment as a result of decommissioning activities and through field use and disposal practices such as open burning. TNT has been found in environmental media only in the vicinity of such sites.

Air. A low vapor pressure, 1.99×10^{-4} mm Hg at 20 °C (HSDB 1995a), indicates a low potential to enter the atmosphere. Nitroaromatics associated with munitions manufacture and processing have generally not been detected in atmospheric monitoring studies (Howard et al. 1976).

Surface Water and Groundwater. Reported levels in production and processing facility effluents vary. Because of color changes when exposed to sunlight, effluents from production and processing facilities are referred to as red or pink water. Spanggord and coworkers (1982a) characterized wastewaters resulting from the production and purification of TNT. They reported that concentrations of 0.10–3.40 mg/L occurred in 20% of condensate wastewater from plants using the sellite manufacturing process. Concentrations in a lagoon receiving pink water ranged from 774 to 998 μg/L (Triegel et al. 1983); concentrations in pink water effluents from LAP plants ranged from 1 to 178 mg/L (Patterson et al. 1977). Concentrations in untreated wastewater at the Radford AAP, VA, ranged from 101 to 143 mg/L (Nay et al. 1972) and up to 120 mg/L at several manufacturing plants (Freeman and Colitti 1982; Andren et al. 1977). Concentrations in aqueous samples collected onsite at the Joliet AAP (Illinois) ranged from <0.2 μg/L (the detection limit) to 1.0 mg/L, the latter in an open wastewater channel (Jerger et al. 1976). Sediment concentrations ranged from <1 mg/kg to 44,200 mg/kg (air dry weight) in a red water pond at the same site. Respective concentrations were lower at the Burlington, IA, AAP. There, sediment concentrations below an inactive lagoon were as high as 300 mg/kg (Sanocki et al. 1976). At the Lone Star AAP, TX, sludge from below pink water settling ponds contained up to 50,000 mg/kg (Phung and Bulot 1981). Concentrations in onsite streams at the Iowa AAP ranged up to 29 μg/L while average concentrations in stream sediment cores ranged up to 111 mg/kg (Sanocki et al. 1976).

Runoff and leaching from lagoons and storage and disposal areas may contaminate surface water and groundwater. However, concentrations downstream of discharge points rapidly decrease because of dilution and photolysis (Spanggord et al. 1980b). TNT has been found in groundwater collected at several munitions facilities. Concentrations up to 350 μg/L were measured in groundwater offsite at the Cornhusker AAP near Grand Island, Nebraska (Spalding and Fulton 1988). TNT concentrations of 320 μg/L (61 m downgradient) and 1 μg/L (326 m downgradient) were measured in groundwater samples at the Hawthorne Naval Ammunition Depot, NV (Goerlitz and Franks 1989). TNT was not detected in ocean waters near ocean dumping sites for waste munitions (Hoffsommer and Rosen 1972) nor in ocean floor sediments near dumping sites (Hoffsommer et al. 1972).

Soils. Detectable concentrations of TNT in soil at the Alabama AAP (Talladega County) ranged from <0.4 to 2.3 mg/kg (Rosenblatt and Small 1981). Areas sampled included a smokeless powder manufacturing area, magazine area, flashing ground, and an aniline sludge basin. At the Joliet AAP, soil concentrations ranged from less than the limit of detection to 87,000 mg/kg, with highest concentrations next to a LAP area (Phillips et al. 1994; Simini et al. 1995). At the U.S. Department of Energy Weldon Springs site in St. Charles County, MO, TNT was detected in surface soil samples at an average concentration of 13,000 mg/kg (Haroun et al. 1990). At the West Virginia Ordinance Works, TNT and other nitroaromatics have been detected in soil at concentrations up to 40,000 mg/kg at burning sites (Kraus et al. 1985). At the Lone Star AAP, TX, surface soil samples at 0.2–0.6 m depth contained 19 mg/kg (Phung and Bulot 1981). In a survey of TNT and transformation products in a limited number of soil samples collected at AAP, depots, and arsenals, the following concentrations (air-dried weight) of TNT were found: Nebraska Ordnance Works, 0.12–20,600 mg/kg; Umatilla Depot, OR, 131–38,600 mg/kg; Newport, IN, 0.4 mg/kg; Weldon Springs Training Area, MO, 0.12–13,400 mg/kg; Iowa AAP, 0.63–15,400 mg/kg; Raritan Arsenal, NJ, 1.19–745 mg/kg; Hawthorne AAP, NV, 4.4–13,900 mg/kg; Hastings East Park, NE, 0.12–10.6 mg/kg; Milan, TN, 1.1–35 mg/kg; Louisiana AAP, 12.4–14.8 mg/kg; VIGO Chemical Plant, IN, 0.56–7.61 mg/kg; Chickasaw Ordnance Works, TN, 0.12 mg/kg; Sangamon Ordnance Plant, IL, 103 mg/kg; Lexington-Bluegrass Depot, KY, 5.90 mg/kg; Eagle River Flats, AK, 0.12–115 mg/kg; and Camp Shelby, MO, not detected (Walsh and Jenkins 1992). Four soil samples taken from the Naval Surface Warfare Center, Crane, IN, contained concentrations ranging from 0.12 to 2.3 mg/kg (dry weight) (Grant et al. 1995). The U.S. Army Installation Restoration Data Management Information System records concentrations up to 711,000 mg/kg at an unidentified site.

2. Transport and Transformation Processes.

Abiotic Processes. No information on the transformation of TNT in air was found. TNT in the atmosphere should undergo direct photolysis as it does in surface water. Based on the rate of photolysis in distilled water, the estimated photolytic half-life in the atmosphere would range from 3.7 to 11.3 hr (Howard et al. 1991). Based on the estimated rate constant for reaction with hydroxyl radicals in the atmosphere, the photooxidation half-life would range from 18.4 to 184 d (Howard et al. 1991).

Photolysis is the primary environmental transformation process in water. Laboratory studies of TNT photodecomposition in water indicate that decomposition occurs much more rapidly in sunlight than in darkness and that the rate is accelerated by the presence of degradation products. In natural waters, the rate of photolysis is accelerated by the presence of natural substances (Spanggord et al. 1980b; Mabey et al. 1983). In laboratory studies, TNT photolyzed rapidly in natural water, with half-lives of 0.5–22 hr; in field studies, the compound rapidly declined within a short distance of discharge points.

Photolysis results in the formation of pink water, which is rich in nitro compounds. Numerous degradation products of TNT have been identified in pink water. Burlinson (1980) proposed a mechanism for photodecomposition of TNT; the major transformation products in natural water are 1,3,5-trinitrobenzene, 4,6-dinitroanthranil, 2,4,6-trinitrobenzaldehyde, 2,4,6-trinitrobenzonitrile, and 2,4,6-trinitrobenzoic acid. In addition, a number of azo and azoxy derivatives formed by the coupling of nitroso and hydroxylamine products have been identified (Burlinson 1980; Jerger et al. 1976; Mabey et al. 1983; Spanggord et al. 1980a). TNT may be more persistent in deep, quiescent waters where sunlight is attenuated. Hydrolysis and volatilization from water are not expected under environmental conditions (HSDB 1995a). TNT was stable in seawater for 108 d under neutral conditions, indicating stability against hydrolysis (Howard et al. 1976). A calculated Henry's law constant of 4.57×10^{-7} atm-m^3/mole at 20 °C indicates extremely slow volatilization from water or moist soils (HSDB 1995a). Water solubility of TNT is 130 mg/L (Ryon 1987).

Numerous studies have also investigated the fate of TNT in soil. Jenkins (1989) exposed TNT-containing soils from two AAPs to available sunlight as well as fluorescent lighting over a period of 10 d. The soils, which were spread in a thin layer, were stirred several times per day. After 10 d, a loss of 10% compared to samples maintained in the dark was measured. Spanggord et al. (1980a) reviewed studies on sorption of TNT on soils and sediments and found sorption values of 5.5 to 19.3. Partition coefficients or K_p ([μg chemical in soil/g of soil]/[μg chemical in water/g of water]) in the following studies range from 4 to 53, indicating little sorption (strongly adsorbed chemicals have K_p values greater than 100).

Soils from 13 AAPs were tested for adsorption and desorption of TNT (Pennington 1988a; Pennington and Patrick 1990). The soils were primarily silt loams, low in organic carbon content (0.3%–3.6%). Both adsorption and desorption reached steady states within 2 hr. With the addition of 320 μg TNT/g of soil, adsorption percentages ranged from 7.5% to 32% (24–102 μg/g). Sequential desorption with water indicated that most of the TNT was desorbed after three sequential desorption cycles. The average percent of TNT desorbed was 88%; i.e., remaining adsorbed TNT averaged 12% (7.7%–23%; 2–24 μg/g). Adsorption was highest in a clay soil (32% initially and 23% following sequential desorption). The average K_p value was 4, but the values varied with soil type (range, 2–11). In another study, K_p values of four aquatic sediments were 5.5, 14.3, 16.5, and 22.2 after 24-hr equilibrium time, also suggesting low sorption (Sikka et al. 1980). Slightly higher values were measured by Spanggord et al. (1980b); the average K_p of five sediments (average organic carbon content, 3.3%) from water bodies receiving munition wastewater discharges was 53 and the average K_{oc} [(μg adsorbed/g organic carbon)/(μg/mL solution)] was 1600. Cataldo et al. (1989) found that the fraction of soil-sorbed ^{14}C-labeled TNT/TNT metabolites increased with time. The amount of label not readily removed by exhaustive extraction increased from <6% at time 0 to 50% after 60 d of incubation.

The transport of pink water compounds including TNT (70 mg/L) through columns of garden soil (6.5% organic matter content) amended with microbes from activated sludge and anaerobic sludge digest was studied by Greene et al. (1984). Pink water solutions were continuously pumped through the columns and effluent samples were collected weekly over 110 d. Flow rates were varied and some columns were amended with glucose. Low recoveries in the leachate from most of the columns and the presence of metabolites indicated that TNT had either remained in the soil or been biotransformed.

Checkai et al. (1993) collected intact soil-core columns from an uncontaminated area at the Milan AAP to study transport and transformation of munitions chemicals in site-specific soils. The soil was a Lexington silt loam with a 15-cm A horizon containing 16 g/kg organic matter and a B horizon extending from 15 to 69 cm and containing 5 g/kg organic matter. A mixture of munitions simulating open burning/open detonation ash was added to the soil surface. Concentrations were 1000 mg/kg TNT, 1000 mg/kg 2,4-dinitrotoluene, and 400 mg/kg 2,6-dinitrotoluene. The columns were leached with simulated rainfall over a period of 32.5 wk; controlled tension was applied. Initially, half (5/10) of the treatment columns yielded leachates containing low concentrations of TNT (0.09–0.21 mg/L). By d 10, TNT was no longer detectable (<0.09 mg/L). No transformation products were detectable in the leachates. Neither TNT nor its transformation products were found in the soil during the first sampling at 6.5 wk.

In a field study, TNT was mixed with soil at a ratio of 1/1000, placed in open-ended tubes buried in the ground, and allowed to weather for 20 yr (DuBois and Baytos 1991). A bacterium (*Pseudomonas aeruginosa*) added to the tubes survived less than 6 mon. The half-life of TNT under these conditions was estimated at 1 yr.

Biotransformation. TNT is biodegraded in water, soil, sediment, and sludge by bacterial and fungal species under both aerobic and anaerobic conditions (Spanggord et al. 1980a); the process occurs more slowly than photolysis, with the estimated half-life ranging from 1 to 6 mon in surface waters (ATSDR 1995a). Transformation occurs faster under anaerobic conditions than under aerobic conditions (Preuss et al. 1993). Reduction of a nitro group is the first step in transformation.

Spanggord et al. (1980b) demonstrated the degradation of TNT in natural waters collected below several AAPs. Degradation in water alone was slow, with half-lives of 19–25 d. Addition of organic nutrients or sediment accelerated the process. Studies with ^{14}C-ring labeled TNT demonstrated that while the nitro groups were reduced, ring cleavage did not occur. The metabolites formed in water, soil, or sediments include 2-amino-4,6-dinitrotoluene, 4-amino-2,6-dinitrotoluene, hydroxyaminodinitrotoluenes, 2,4-diamino-6-nitrotoluene, 2,6-diamino-4-nitrotoluene, 2,4,6-triaminotoluene, and tetranitroazotoluene compounds (Fig. 1) (Burlinson 1980; Carpenter et al. 1978; Hoffsommer et al. 1978; Jerger et al. 1976; McCormick et al. 1976; Parrish 1977; Spain 1995; Spanggord et al. 1980b; Weitzel et al. 1975; Won et al. 1974). The formation of azoxy compounds probably results from the condensation of hydroxylamino and C-nitroso

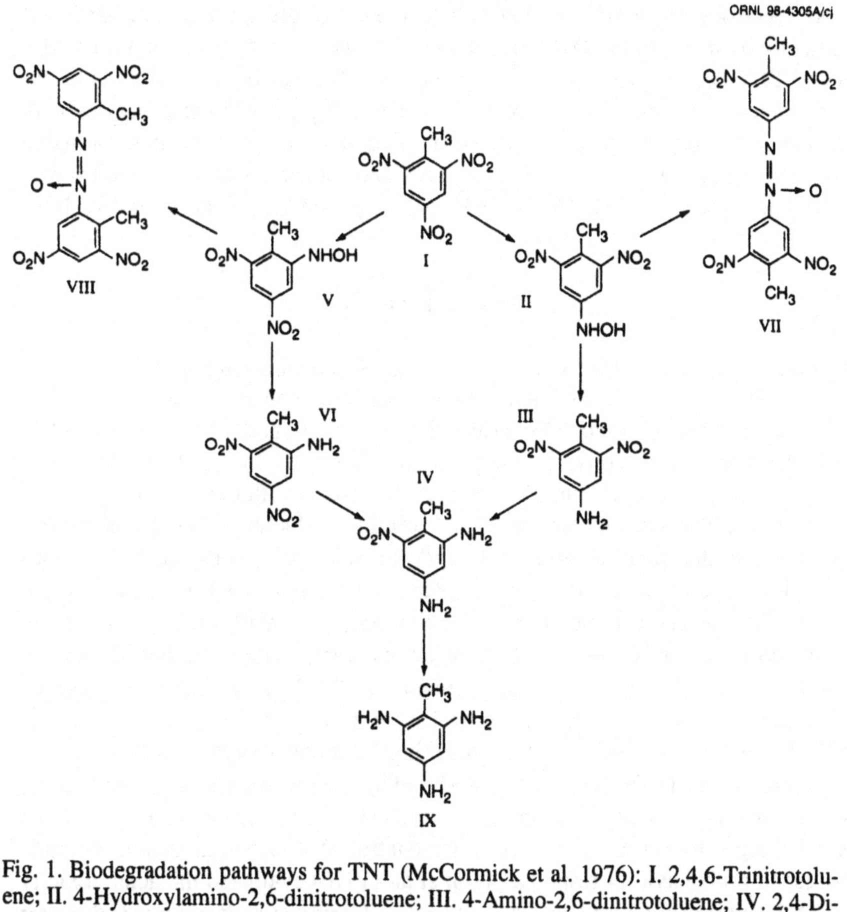

Fig. 1. Biodegradation pathways for TNT (McCormick et al. 1976): I. 2,4,6-Trinitrotoluene; II. 4-Hydroxylamino-2,6-dinitrotoluene; III. 4-Amino-2,6-dinitrotoluene; IV. 2,4-Diamino-6-nitrotoluene; V. 2-hydroxylamino-4,6-dinitrotoluene; VI. 2-Amino-4,6-dinitrotoluene; VII. 2,2′,6,6′-tetranitro-4,4′azoxytoluene; VIII. 4,4′6,6′-tetranitro-2,2′-azoxytoluene; IX. 2,4,6-triaminotoluene

compounds. Although complete mineralization did not occur in most of these studies, ring degradation of ^{14}C-ring-labeled TNT was reported under special conditions. For example, it occurred in tests using the white-rot fungus, *Phanerochaete chrysosporium* (Fernando et al. 1990), under carefully controlled anaerobic conditions (Funk et al. 1993), and following removal of the nitro groups (Marvin-Sikkema and de Bout 1994). Under anaerobic conditions, 2,4,6-trihydroxytoluene and *p*-hydroxytoluene and several unidentified intermediates were present (Funk et al. 1993). Phloroglucinol and pyrogallol were reported as products of 2,4-diamino-6-nitrotoluene degradation under aerobic conditions (Naumova et al. 1988).

Cataldo et al. (1989) amended various soils with unlabeled or ^{14}C-labeled-TNT and isolated the degradation products. The soils ranged from sandy to silt loams with organic carbon contents of 0.5%–7.2%. The isomers 2-amino-4,6-dinitrotoluene and 4-amino-2,6-dinitrotoluene (as determined by HPLC and mass spectrometry) appeared almost immediately in the soil and increased dramatically by d 10. After 60 d of incubation the two isomers accounted for >80% of the TNT-derived activity in two of the soils. Similar but slower transformation occurred in radiation-sterilized soils, indicating that abiotic processes may account for some or all of the transformation. The 4-amino-2,6-dinitrotoluene isomer was present at nearly twice the abundance of the 2-amino-4,6-dinitrotoluene isomer (Harvey et al. 1990). In Palouse soil, 4% of the TNT was oxidized to $^{14}CO_2$.

B. Aquatic Toxicology

1. Invertebrates and Fish: Acute Effects. Studies of the acute toxicity of TNT, conducted on five species of invertebrates and four species of fish, are reported in Table 5. These tests typically lasted 48 hr for invertebrates (except the rotifer) and 96 hr for fish; the reported endpoint was mortality (LC_{50} values). Test temperatures ranged from 20 ° to 25 °C except for studies with rainbow trout (10 °C) or where the effect of temperature was determined. Median lethal concentrations for invertebrates ranged from >4.4 mg/L to >29 mg/L; for fish, LC_{50} values ranged from 0.8 mg/L to 3.7 mg/L. Pederson (1970) and Liu et al. (1983b) found that the acute toxicity of TNT was only minimally affected by water temperature, hardness, and pH. Both acute and some chronic studies were complicated by the photochemical transformation of TNT and the resulting presence of degradation products. According to Liu et al. (1983b), extensive photolysis of TNT reduced the toxicity of the resulting solutions to aquatic organisms.

Smock et al. (1976) reported a 96-hr EC_{50} of 0.46 mg/L for behavioral response of bluegill sunfish. Responses consisted of shock (gasping at the surface and lethargy) followed by loss of motor control. A concentration of 0.05 mg/L produced no behavioral response. Won et al. (1976) evaluated the acute toxicity of TNT to two saltwater species, a copepod, *Tigriopus californicus*, and oyster larvae, *Crassostrea gigas*. Groups of 100 copepods and 200 oyster larvae (average age, 20 d) were incubated at 20 °C for 72 and 96 hr, respectively, and

Table 5. Acute toxicity of TNT to aquatic invertebrates and fish.

Test species	Test duration	LC_{50} (mg/L)	Reference
Brachionus calyciflorus (rotifer)	24-hr	9.1[a]	Snell and Moffat 1992
Brachionus calyciflorus (rotifer)	24-hr	5.6[a]	Toussaint et al. 1995
Lumbricus variegatus (oligochaete)	48-hr	5.2[a]	Liu et al. 1983b
	48-hr	>29.0[b]	
Daphnia magna (water flea)	48-hr	11.7[a]	Liu et al. 1983a; Liu et al. 1983b
	48-hr	>4.4[b]	
Hyallela azteca (scud)	48-hr	6.5[a]	Liu et al. 1983b
Tanytarsus dissimilis (midge)	48-hr	27.0[a]	Liu et al. 1983b
Ictalurus punctatus (channel catfish)	96-hr	2.4[a]	Liu et al. 1983a
		3.3[b]	Liu et al. 1983b
Lepomis macrochirus (bluegill sunfish)	96-hr	2.3[a,c]	Pederson 1970
	96-hr	2.3[a,d]	Pederson 1970
	96-hr	2.7[a,e]	Pederson 1970
	96-hr	2.8[a,f]	Pederson 1970
Lepomis macrochirus (bluegill sunfish)	96-hr	2.6[a]	Nay et al. 1974
Lepomis macrochirus (bluegill sunfish)	96-hr	2.6[a]	Liu et al. 1983a
	96-hr	3.4[a]	Liu et al. 1983a
	96-hr	2.5[b]	Liu et al. 1983b
Pimephales promelas (fathead minnow)	96-hr	2.58[b]	Smock et al. 1976
Pimephales promelas (fathead minnow)	96-hr	2.9[a]	Liu et al. 1983a
	96-hr	3.0[a]	Liu et al. 1983b
	96-hr	3.7[b]	
Pimephales promelas (fathead minnow)	96-hr	3.0[a]	Bailey and Spanggord 1983
Oncorhynchus mykiss (rainbow trout)	96-hr	1.5[a]	Liu et al. 1983b
	96-hr	0.8[a]	
	96-hr	2.0[b]	

[a]Static tests.
[b]Flow-through tests.
[c]25 °C, 60 ppm $CaCO_3$.
[d]25 °C, 180 ppm $CaCO_3$.
[e]10 °C, 60 ppm $CaCO_3$.
[f]10 °C, 180 ppm $CaCO_3$.

evaluated for mortality. The lowest concentration resulting in death of copepods was 2.5 mg/L (18% mortality); mortality was 44% at 5 mg/L. For oyster larvae, 5 mg/L resulted in no deaths, but 10 mg/L was toxic, resulting in a mortality of 82%. LC_{50} values were not computed.

2. Invertebrates and Fish: Chronic Effects. Chronic tests, conducted with two species of invertebrates and three species of fish, are reported in Table 6. The 48-hr test with rotifers (Snell and Moffat 1992) can be considered a chronic test because several generations were produced within the time period. Although EPA-recommended flow-through conditions were not used in the study with rotifers, incubation occurred in the dark, limiting the photolysis of TNT. The lowest concentration that statistically significantly affected reproduction of the rotifer during 48 hr was 5.0 mg/L. However, the EC_{50} for reproduction estimated from a linear regression was 4.0 mg/L.

In a 28-d study, *Daphnia magna* was tolerant of TNT at ≤0.48 mg/L. Using flow-through conditions, Bailey et al. (1985) tested the chronic effects of TNT at concentrations of 0, 0.03, 0.08, 0.24, 0.48, and 1.03 mg/L on *D. magna*. Daphnids were observed for mortality, reproductive success at 14, 21, and 28 d, onset of reproduction, and body length of young. These concentrations had no effect on cumulative mortality, time to first brood, or length. Significant decreases occurred in the number of young produced at d 14 and 21 at the 1.03 mg/L concentration in one series, but not by d 28. The authors considered this a transitory effect, but because average daphnid survival in the environment is less than 28 d, they thought that the impaired reproduction could have biological significance and thus considered this an effect level. The no-effect level was 0.48 mg/L.

Bailey et al. (1985) investigated the effects of TNT on early life stages of channel catfish, rainbow trout, and fathead minnows. In 30- or 60-d studies, the lowest effective concentration was 0.24 mg/L, which resulted in decreased survival of rainbow trout fry at 60 d. However, in a two-generation study, a concentration as low as 0.04 mg/L decreased several reproductive parameters of F_0 fathead minnows. Groups of channel catfish eggs were exposed in duplicate flow-through chambers to 0.0, 0.11, 0.15, 0.30, 0.66, or 1.35 mg/L; the temperature was 25 °C and the photoperiod was 16 hr light:8 hr dark. Following hatch, the fry were exposed in the chambers for a 30-d period. Groups of 30 eggs of fathead minnows were exposed under similar conditions (with a variable photoperiod) to 0.0, 0.07, 0.10, 0.16, 0.42, or 0.84 mg/L for a total of 30 d. Fry length was also measured. Groups of 60 eggs of rainbow trout were exposed at 12 °C to 0.0, 0.07, 0.12, 0.21, 0.49, or 0.93 mg/L (through 30 d post hatch) or 0.0, 0.02, 0.04, 0.13, 0.24, 0.50, 0.87, or 1.69 mg/L (through 60 d post hatch). Fry were weighed and observed for deformities in the second test.

Concentrations of 1.35 mg/L or less had no effect on percent hatch of eggs or survival of channel catfish fry over a 30-d period. In tests with rainbow trout, hatch was decreased at 0.93 mg/L in the first test, but in the second test, hatch was not affected by ≤1.69 mg/L. Concentrations resulting in reduced survival

Table 6. Chronic toxicity of TNT to aquatic invertebrates and fish[a].

Species	Stage/age	Parameter measured	Parameter response[b]	Concentration (mg/L)	Reference
Brachionus calyciflorus (rotifer)	Newly hatched	Reproduction (48-hr life cycle test)	NOEC	2.3	Snell and Moffat 1992
			LOEC	5.0	
			EC_{50}	4.0	
Daphnia magna (water flea)	Young	Mortality (28 d)	No effect	0.03–1.03	Bailey et al. 1985
		Total reproduction	No effect	0.48	
			Effect, 14, 21 d	1.03	
		Young/female	No effect	0.03–1.03	
		Young/d/female	No effect	0.03–1.03	
		Length	No effect	0.03–1.03	
		Days, first brood	No effect	0.03–1.03	
Ictalurus punctatus (channel catfish)	Eggs, fry	Percent hatch	No effect	0.11–1.35	Bailey et al. 1985
		Fry survival (30 d)	No effect	0.11–1.35	
Oncorhynchus mykiss (rainbow trout)	Eggs, fry	Test 1:			Bailey et al. 1985
		Number hatched	Decreased hatch	0.93	
		Fry survival (30 d)	Decreased	0.93	
		Average length	Decreased	0.49	
		Average weight	Decreased	0.49	
		Test 2:			
		Number hatched	No effect	0.02–1.69	
		Fry survival (60 d)	Reduced	0.24	
		Average length	Decreased	0.50	
		Average weight	Decreased	0.87	
		Deformities	No effect	0.02–1.69	

Pimephales promelas (fathead minnow)	Eggs, fry	Percent hatch	No effect	0.07–0.84	Bailey et al. 1985
		Fry survival (30 d)	Reduced	0.84	
		Average length	Inconclusive	0.84	
Pimephales promelas (fathead minnow)	F_0 and F_1 generations	F_0 eggs:			Bailey et al. 1985
		Number hatched	Reduced	1.21 (in one series)	
		Number fry deformed	No effect	0.04–1.21	
		Fry survival	Reduced, 30–178 d	0.25 (in one series)	
		Fry length	Inconclusive		
		F_0 reproduction:			
		Pair survival	Decreased	0.04–1.21	
		No. of spawns/pair	Decreased	0.04–1.21	
		Eggs/pair	Decreased	0.04–1.21	
		Eggs/spawn	Decreased	1.21	
		Eggs/pair/d	Decreased	0.56	
		F_1 eggs:			
		Number hatched	Decreased	0.04–1.21	
		Number deformed	Increased	0.10–1.21	
		Fry survival (30 d)	Decreased	0.04–1.21	
		Fry length (30 d)	No effect	0.04–0.25[c]	
		Fry weight (30 d)	No effect	0.04–0.25[c]	
		Fry survival (60 d)	No effect	0.04–0.25[c]	
		Fry length (60 d)	Decreased	0.04–0.56[c]	
		Fry weight (60 d)	Decreased	0.04–0.56[c]	

NOEC, no-observed-effects concentration; LOEC, lowest-observed-effects concentration;
[a]Test conditions: static (rotifer), flow-through (fish).
[b]Statistically significant responses reported.
[c]No survival above this concentration.

and decreased length and weight were 0.24 and 0.50 mg/L, respectively. There was no increase in deformities at the tested concentrations.

Although percent hatch of fathead minnow eggs was not affected by ≤0.84 mg/L, fry survival was reduced at 30 d for the 0.84 mg/L treatment (50% for controls vs. 32% for treated group) and fry length was reduced in one of the two duplicate chambers at 0.84 mg/L. Statistical analyses were not performed. In comparing the 30-d studies, rainbow trout and fathead minnows appear more sensitive to TNT exposure than channel catfish.

Bailey et al. (1985) also studied effects of chronic exposure of fathead minnows to 0.00, 0.04, 0.10, 0.25, 0.56, or 1.21 mg/L of TNT over two generations. The F_1 generation was observed for 30 or 60 d post hatch under test conditions. F_0 and F_1 egg and fry survival, length, deformities, and F_0 fertility indices were determined. Most reproductive parameters of breeding pairs that developed from the F_0 eggs were adversely affected at all concentrations. The overall effect on reproduction was reduction in survival of breeding pairs and in frequency of spawns per pair. Total survivability and total productivity indices indicated that TNT had a deleterious effect on fathead minnows at all concentrations.

Incipient LC_{50} values, the concentration above which 50% of organisms cannot survive indefinitely, were determined for several aquatic species (Liu et al. 1983b). The tests were conducted under flow-through conditions for various periods of time. Incipient LC_{50} values and exposure time in hours (in parentheses) were *D. magna*, 0.19 mg/L (192); *Lumbricus variegatus*, 13.9 mg/L (336); *Lepomis macrochirus*, 1.4 mg/L (312); *Oncorhynchus mykiss*, 1.9 mg/L (240); *Ictalurus punctatus*, 1.6 mg/L (288), and *Pimephales promelas*, 1.5 mg/L (384). These values for lethality are higher than for the longer-term studies in which reproduction was the endpoint.

3. Plants. Short-term and chronic tests using growth parameters as the endpoint were conducted with several species of algae and an aquatic vascular plant (Table 7). As noted by some of the authors, the light necessary for plant growth in these static tests resulted in photolysis of the TNT. Fitzgerald et al. (1952) exposed cultures of the blue-green alga, *Microcystis aeruginosa*, grown in 125-mL flasks at 22 °C to TNT for 24 hr. A concentration of 8 mg/L resulted in 100% mortality. Bringmann and Kuhn (1980) tested *Scenedesmus quadricauda* colonies for effects using the cell multiplication inhibition test. Colonies were exposed to a series of TNT concentrations for 16 hr at 27 °C; effects were measured by comparing turbidimetric values expressed by the extinction of primary light of monochromatic radiation at 578 nm for a 10-mm-thick layer. The "toxicity threshold" was 1.6 mg/L.

In longer-term studies, Won et al. (1976) evaluated the toxicity of TNT to the green alga *Selenastrum capricornutum*. Incubation was for 7 d and toxicity was evaluated in terms of dry weight and morphological alterations. A concentration of 2.5 mg/L significantly depressed growth (44% of controls) and produced a population of ballooned, extensively granulated cells. A chronic EC_{50} was not calculated.

Table 7. Toxicity of TNT to algae and a vascular plant[a].

Test species	Test duration	Effect	Concentration (mg/L)	Reference
Microcystis aeruginosa (blue-green alga)	24 hr	100% mortality	8	Fitzgerald et al. 1952
Microcystis aeruginosa (blue-green alga)	14 d	Growth inhibition (LOEC)	4.1	Liu et al. 1983b
Microcystis aeruginosa (blue-green alga)	15 d	No effect on growth Initial decrease in growth Permanent decrease in growth	15 25 50	Smock et al. 1976
Anabaena flos-aquae (blue-green alga)	14 d	Growth inhibition (LOEC)	8.2	Liu et al. 1983b
Scenedesmus quadricauda (green alga)	16 hr	"Toxicity threshold"	1.6	Bringmann and Kuhn 1980
Selenastrum capricornutum (green alga)	96 hr	EC_{50} for growth	1.1	Sunahara et al. 1995
Selenastrum capricornutum (green alga)	7 d	Growth depression (12%) Growth depression (56%); changes in morphology	1.0 2.5	Won et al. 1976
Selenastrum capricornutum (green alga)	14 d	EC_{50} for growth	1.5	Bailey 1982a
Selenastrum capricornutum (green alga)	14 d	Growth inhibition (LOEC)	4.1	Liu et al. 1983b
Selenastrum capricornutum (green alga)	17 d	No effect on growth Initial decrease in growth Significant decrease in growth	3 5 ≥7	Smock et al. 1976
Navicula pelliculosa (diatom)	14 d	Growth inhibition (LOEC)	18	Liu et al. 1983b
Lemna perpusilla (vascular plant)	11 d	No effect on colony growth Growth depression (≥10%) Mortality	0.5 1.0 ≥5	Schott and Worthley 1974

[a]All tests were conducted under static conditions.

Smock et al. (1976) exposed cultures of *S. capricornutum* to 0, 1, 3, 5, 7, or 9 mg/L for 17 d and cultures of *M. aeruginosa* to 0, 5, 10, 15, 25, or 50 mg/L for 15 d under static test conditions. Concentrations up to 3 mg/L had no effect on the growth of *S. capricornutum*. An initial decrease in growth occurred at concentrations ≥5 mg/L, but the effect disappeared by d 17. For *M. aeruginosa*, no effects on growth occurred up to 15 mg/L, temporary decreases occurred at 25 mg/L, and growth was permanently retarded throughout the 15-d assay at 50 mg/L. This result contrasts with that of Fitzgerald et al. (1952), who reported 100% mortality for *M. aeruginosa* at 8 mg/L. Smock et al. (1976) reported that TNT could no longer be measured in the test solutions after 7 d.

Liu et al. (1983b) studied the effects of TNT on several species of algae—*S. capricornutum, M. aeruginosa, Anabaena flos-aquae* (blue-green alga), and *Navicula pelliculosa* (diatom) using the USEPA bottle technique (static test). Effects were detected by measuring cell concentrations following exposure to a series of TNT concentrations for 14 d. Growth stimulation occurred at low concentrations. The lowest concentrations that inhibited growth were 4.1 mg/L for *S. capricornutum* and *M. aeruginosa*, 8.2 mg/L for *A. flos-aquae*, and 18 mg/L for *N. pelliculosa*. However, photolysis of the TNT, as indicated by the reddish-brown color of the solutions, made the results of these tests unreliable. In a more recent study, the 96-hr EC_{50} for growth of *S. capricornutum* was 1.1 mg/L Sunahara et al. 1995). Details of the study were not given.

Schott and Worthley (1974) exposed the aquatic vascular plant *Lemna perpusilla* to concentrations of TNT ranging from 0.01 to 50 mg/L for 11 d. A concentration of 0.5 mg/L had no effect on colony growth, whereas growth was depressed at 1.0 mg/L and death occurred at ≥5.0 mg/L. A change in pH (6.3 or 8.5) did not affect toxicity.

4. Metabolism and Bioconcentration. Studies on the metabolism of TNT by aquatic organisms were not located in the available literature. Calculated and measured log octanol-water partition coefficients (log K_{ow}) values of 1.6–2.7 (ATSDR 1995a) indicate a low potential for bioconcentration.

The potential for TNT to bioconcentrate in aquatic species as determined by uptake of ^{14}C-labeled TNT was evaluated by Liu et al. (1983a,b). Tests were performed for 96 hr under static exposure conditions. The 4-d bioconcentration factor (BCF) was calculated by dividing the average amount of radioactivity in the biological samples by the amount in the test water; no corrections were made for metabolites. For intact organisms, the BCF ranged from 202 to 453 (Table 8). The authors considered that steady-state conditions were not achieved; thus, the values represent one point on the uptake curve. Because steady-state conditions were not reached and metabolic products were not considered, Liu et al. (1983a,b) also calculated the potential to bioconcentrate in fish using a calculated log K_{ow} (2.03) and the relationship between log K_{ow} and the steady-state BCF developed by Veith et al. (1980). The calculated value BCF of 20.5 represents a low potential for bioconcentration. Although the implication of not

Table 8. Bioconcentration factors (BCF) of TNT in aquatic species.

Species	Value	Comments	Reference
Fish	20.5	Steady-state BCF computed using calculated log K_{ow} of 2.03 and equation for organism lipid content of 8%	Liu et al. 1983a Liu et al. 1983b
Selenastrum capricornutum (green alga)	453	4-d BCF, static test	Liu et al. 1983b
Daphnia magna (water flea)	209	4-d BCF, static test	Liu et al. 1983b
Lumbricus variegatus (oligochaete)	202	4-d BCF, static test	Liu et al. 1983b
Lepomis macrochirus (bluegill)		4-d BCF, static test	Liu et al. 1983b
Viscera	338		
Muscle	9.5		

reaching a steady state in the laboratory study would be empirical BCF values that are too low, it should be noted that the measured values are much higher than the modeled value.

C. Aquatic Criteria and Screening Benchmarks

1. Aquatic Organisms. Although the USEPA has not calculated numerical national WQC for TNT, sufficient data are available to do so according to the guidelines (Stephan et al. 1985). Data are sufficient for deriving both acute and chronic criteria. Acceptable acute toxicity test results were available for eight different families: (1) Salmonidae (*Oncorhynchus mykiss*), (2) a second family in the class Osteichthyes (*Lepomis macrochirus*), (3) a third family in the phylum Chordata (*Ictalurus punctatus* or *Pimephales promelas*), (4) a planktonic crustacean (*Daphnia magna*), (5) a benthic crustacean (*Hyallela azteca*), (6) an insect (*Tanytarsus dissimilis*), (7) a family in a phylum other than Arthropoda (*Lumbricus variegatus*), and (8) a family in any order of insect or any phylum not already represented (*Brachionus calyciflorus*).

Genus mean acute values were calculated using results from both the static and flow-through tests in Table 5 because the values were very similar. The nine GMAVs were ordered and the four lowest were assigned ranks (*R*) from the lowest to highest: rainbow trout, Rank 1; bluegill sunfish, Rank 2; channel catfish, Rank 3; and fathead minnow, Rank 4 (Table 9). The tests with bluegill sunfish conducted at 10 °C were not included in the GMAV calculation. From these data a FAV was calculated.

Table 9. Ranking of acute aquatic toxicity studies for TNT.

Rank (R)	GMAV[a]	ln GMAV[b]	$(\ln \text{GMAV})^2$	$P = R/(N+1)$[c]	$\sqrt{P}$
4	3.02	1.1053	1.2216	0.4	0.6325
3	2.81	1.0332	1.0675	0.3	0.5477
2	2.59	0.9517	0.9057	0.2	0.4472
1	1.34	0.2927	0.0857	0.1	0.3162
Sum		3.3828	3.2804	1.0	1.9436

[a]GMAV, genus mean acute value of LC_{50} values in mg/L, based on data in Table 5.
[b]ln GMAV, natural log of GMAV.
[c]P, comparative probability for each GMAV; N, 9 GMAVs.

$$S^2 = \frac{\Sigma[(\ln \text{GMAV})^2] - [\frac{(\Sigma \ln \text{GMAV})^2}{4}]}{\Sigma(\text{P}) - \frac{[(\Sigma\sqrt{\text{P}})^2]}{4}} = 7.55$$

$$S = 2.75$$

$$\text{L} = \frac{[\Sigma(\ln \text{GMAV}) - \text{S}(\Sigma\sqrt{\text{P}}))]}{4} = -0.4893$$

$$\text{A} = \text{S}(\sqrt{0.05}] + \text{L} = 0.1250$$

$$\text{FAV} = \text{e}^{0.1250} = 1.1332 \text{ mg/L}$$

The CMC is one-half of the FAV or 0.57 mg/L.

The FCV was calculated by dividing the FAV by the FACR. Because data from chronic toxicity tests using eight families were not available, the chronic criterion was derived from three or more acute/chronic ratios (ACR). Using the data in Tables 5 and 6, ACR were available for a rotifer, a daphnid, rainbow trout, and the fathead minnow (Table 10). The chronic values are geometric means of LOECs and NOECs. With the exception of the rotifer, all chronic tests were performed by the same investigator under flow-through conditions, and several of the acute and chronic tests used to calculate ACR were performed by the same investigator. Because a lowest-effect concentration was not attained in the tests with the channel catfish, this species was not used. On the other hand, a clear no-effect concentration was not attained for the fathead minnow, but to consider a sensitive species, the LOEC of 0.04 mg/L from the two-generation study was used in the calculations. The FACR is the geometric mean of 2.36, 16.6, 7.59, and 75.5 or 12.2.

$$\text{FCV} = \frac{\text{FAV}}{\text{FACR}} = \frac{1.13}{12.2} = 0.093 \text{ mg/L}$$

Table 10. Aquatic acute/chronic ratios (ACR) for TNT.

Species	Acute values (mg/L)	Chronic values (mg/L)	Acute/chronic ratios (ACR)
Rotifer	9.1	3.0	3.00
	5.6	3.0	1.85
			Geometric mean, 2.36
Daphnid	11.7	0.703	16.6
Channel catfish	2.81	1.35 (no effect)	2.08
Rainbow trout	1.34	0.177	7.59
Fathead minnow	3.02	0.04 (effect)	75.5

Three of the ACRs are within a factor of approximately 10 of each other; the ACR for the fathead minnow was higher than the other values, but was included to be conservative. The test with the fathead minnow was a full life cycle test and indicates the sensitivity of fish reproduction and development to TNT. Thus, the lowest-chronic-effect value for fish (0.04 mg/L) may be a better screening benchmark than the FCV or chronic WQC and can be used until a no-effect value is established. Aquatic screening benchmarks are listed in Table 11. The LOEC of 1.0 mg/L for growth of *S. capricornutum* is also listed in Table 11.

2. Sediment-Associated Organisms. No laboratory studies utilizing benthic organisms and sediments were located. In a survey of periphyton and benthic macroinvertebrate communities in streams at the Iowa AAP, no correlations could be made between aqueous and sediment levels of TNT and diversity indices (Sanocki et al. 1976). Concentrations in on-site streams ranged up to 29 µg/L while average concentrations in stream sediment cores ranged up to 111 mg/kg. Because no experimental studies were available, the SQB was calculated using the EqP approach.

USEPA suggests using a reliable K_{ow} for calculation of SQB using the EqP method. Because K_{ow} values are available, a K_{oc} can be calculated and a SQB_{oc}

Table 11. Water quality criteria/screening benchmarks for TNT[a].

Criterion	Value
Acute water quality criterion	0.57 mg/L[a]
Chronic water quality criterion	0.09 mg/L[a]
Lowest chronic value, fish	0.04 mg/L
Lowest chronic value, daphnids	1.03 mg/L
Lowest chronic value, plants	1.0 mg/L
Sediment quality benchmark (SQB_{oc})	9.2 mg/kg_{oc}[b]

[a]Calculated by ORNL.
[b]mg TNT/kg organic carbon in the sediment.

can be determined. Measured and calculated log K_{ow} values (see Table 4) are within a small range of each other. The calculated log K_{ow} value of 2.03 (Liu et al. 1983a) is close to the geometric mean of the values. Using the relationship between K_{ow} and K_{oc}, i.e., ($\log_{10}[K_{oc}] = 0.00028 + 0.983 \log_{10}[K_{ow}]$), the K_{oc} is 99. The FCV as calculated here is 0.093 mg/L. The organic carbon-normalized SQB, i.e., mg/kg organic carbon (mg/kg_{oc}), is

$$SQB_{oc} = K_{oc} \times FCV$$
$$SQB_{oc} = 99 \text{ L/kg} \times 0.093 \text{ mg/L} = 9.2 \text{ mg/kg}_{oc}$$

If an organic carbon content in the soil of 1% is assumed, the $SQB_{1\%}$ is calculated as:

$$SQB_{1\%} = 0.01 \times 99 \text{ L/kg} \times 0.093 \text{ mg/L} = 0.09 \text{ mg/kg}$$

D. Terrestrial Toxicology

1. Mammals. No subchronic or chronic studies on the toxicity of TNT utilizing mammalian wildlife were located. Laboratory studies utilizing laboratory animal species were summarized in the companion document to this one, *Toxicity Summary for 2,4,6-Trinitrotoluene* (Opresko 1995a). In the first study, Dilley et al. (1982) administered TNT to dogs, rats, and mice for 13 wk. Groups of 2 male and 2 female beagle dogs were administered doses of 0, 0.2, 2.0, or 20 mg/kg/d in gelatin capsules. Additional treated groups were killed at 4 wk and at 17 wk (following a 4-wk recovery period). The animals were observed once daily and weighed weekly, and food consumption was measured daily. Before, during, and at sacrifice, blood samples were taken for hematology and clinical chemistry parameters. At death, major organs were weighed and tissues and organs were examined microscopically.

Clinical signs and toxic symptoms appeared in dogs administered 20 mg/kg/d. In this group, clinical signs of loose stools, diarrhea, and orange-colored urine were observed. One moribund male was sacrificed at wk 12. At the 13-wk sacrifice, males had significantly lower body weights and significantly increased liver, adrenal, and spleen (accompanied by hemosiderosis) weights. Males and females in the high-dose group had significantly decreased mean corpuscular hemoglobin concentrations, and females in this group had a lower serum glutamic pyruvic transaminase level. Anemia was present during the treatment period. Effects were minor in dogs administered 2.0 mg/kg/d (Table 12).

Groups of five male and five female Swiss-Webster mice were administered 0, 0.001%, 0.005%, 0.025%, or 0.125% TNT in their diet for 13 wk (Dilley et al. 1982). The respective calculated doses were 1.56, 7.46, 35.7, and 193 mg/kg/d for males and 1.57, 8.06, 37.8, and 188 mg/kg/d for females. The protocol was the same as in the same study using dogs. Body weight gains were not affected by treatment. Spleens of mice of both sexes in the highest treatment group were enlarged (statistically significant), and enlargement was accompanied by hemosiderosis. Livers of male mice treated for 13 wk and allowed a 4-

Table 12. TNT toxicity data for mammalian species.

Species	Route	Exposure period	NOAEL (mg/kg/d)	LOAEL (mg/kg/d)	Effect/endpoint	Reference
Rat	Diet	24 mon	0.4	2	Kidney, bone; spleen	Furedi et al. 1984a
	Diet	13 wk	1.4	7	Anemia	Dilley et al. 1982
	Diet	13 wk	5	25	Anemia in males	Levine et al. 1984
	Diet	13 wk	5	25	Testicular atrophy	Levine et al. 1984
	Diet	13 wk	34.7	160	Testicular atrophy	Dilley et al. 1982
Mouse	Diet	13 wk	7.5 (m)	35.7 (m)	Anemia	Dilley et al. 1982
			8 (f)	37.8 (f)		
	Diet	24 mon	10	70	Liver; anemia	Furedi et al. 1984b
Dog	Diet	25 wk	—	0.5	Liver	Levine et al. 1990
	Diet	13 wk	0.2	2	Liver; anemia	Dilley et al. 1982

wk recovery period were enlarged, and evidence of necrosis was present in 2/5 mice. Alterations in hematology values occurred at the 4-wk observation, primarily in the high-dose treatment group, but the changes were not statistically significant at 13 wk (see Table 12).

In the same study (Dilley et al. 1982), groups of five male and five female Sprague-Dawley rats were administered 0. 0.002%, 0.01%, 0.05%, or 0.25% TNT in the diet for 13 wk. Doses calculated by the authors were 0, 1.40, 6.97, 34.7, and 160 mg/kg/d for males and 0, 1.45, 7.41, 36.4, and 164 mg/kg/d for females. Rats receiving the highest dose exhibited anemia, with reduced erythrocytes, hemoglobin, and hematocrit. Body weight gains for both sexes were significantly depressed in this treatment group; spleen weights were significantly increased and testes weights were significantly decreased compared with controls. Histopathological examinations revealed hemosiderosis of the spleen and atrophy of the testes. Testicular atrophy was still present in a group of male rats allowed a 4-wk recovery period (see Table 12).

Levine et al. (1981, 1984) conducted a 13-wk study in which groups of 10 male and 10 female Fischer-344 rats were administered TNT (99% pure) in the diet at 0, 1, 5, 25, 125, or 300 mg/kg/d. Thirty animals served as controls and received rodent chow alone. A significant reduction in weight gain occurred in animals receiving 125 and 300 mg/kg/d. At ≥25 mg/kg/d, food intake was reduced and hypercholesterolemia and anemia occurred. Enlarged spleens with congestion and hemosiderosis, enlarged livers with hepatocellular hypertrophy, testicular atrophy with degeneration of the seminiferous tubular epithelium, and slight increases in kidney weight with deposition of pigment were observed in rats receiving 125 or 300 mg/kg/d. Elevated methemoglobin levels and cerebellar lesions were observed only at the 300 mg/kg/d dose level.

In a 6-mon study, groups of six male and six female beagle dogs were administered TNT by capsule at doses of 0, 0.5, 2, 8, or 32 mg/kg/d (Levine et al. 1990). Toxicological endpoints included clinical signs, body weights, food consumption, clinical biochemistry, hematology, urinalyses, organ weights, and gross and tissue morphology. The high dose of 32 mg/kg/d was lethal to two female dogs. TNT was toxic to the liver as evidenced by hepatocytic cloudy swelling and hepatocytomegaly at all doses, with lesions at the 0.5 mg/kg/d dose being trace to mild. Hemolytic anemia, methemoglobinemia, and splenomegaly with accompanying histological lesions were also observed.

In the first chronic study, groups of 75 male and 75 female Fischer-344 rats were administered TNT in the diet at 0.0, 0.4, 2, 10, or 50 mg/kg/d for up to 24 mon (Furedi et al. 1984a). Dose levels for this study were selected on the basis of results in the aforementioned 13-wk study (Levine et al. 1981). Ten rats/sex/dose were killed at 6 and 12 mon and surviving animals at 24 mon. Survival rates were not altered at 24 mon, but decreases in body weight occurred at 10 and 50 mg/kg/d. The major toxic effects observed at 24 mon were anemia with secondary splenic lesions, hepatotoxicity, and urogenital lesions. Hyperplastic or neoplastic lesions of the liver, kidneys, and urinary bladder were also observed. These effects were seen at doses of 10 and 50 mg/kg/d. Some organ weights were increased, including testes weights, at the interim sacrifices but not at study termination. Splenic congestion, increased amounts of pigment deposition in the kidneys, and bone marrow fibrosis were observed at doses ≥2 mg/kg/d.

In the second chronic study, groups of 75 male and 75 female $B6C3F_1$ mice were administered TNT in the diet at doses of 0.0, 1.5, 10, or 70 mg/kg/d for up to 24 mon (Furedi et al. 1984b). Ten mice/sex/group were sacrificed at 6 and 12 mon. Survival rates were not altered at 24 mon, although mice administered 10 and 70 mg/kg/d had reduced body weight gains. The major systemic effects were a mild anemia for both sexes receiving 70 mg/kg/d and probable hepatotoxicity as indicated by enzyme level changes. Leukemia and malignant lymphoma of the spleen were present in females receiving 70 mg/kg/d. Some organ weight changes were observed (e.g., liver and brain) but were not supported histologically. Testes weight and microscopic appearance were unaltered.

2. Birds. No subchronic or chronic studies on TNT toxicity to birds were located.

3. Plants. Tubers of yellow nutsedge (*Cyperus esculentus*) were germinated in hydroponic solutions containing 0, 5, 10, or 20 mg/L TNT (Palazzo and Leggett 1986). Solutions were replaced after 3 wk and plants were harvested after 42 d. Leaf and root growth were significantly reduced at all TNT concentrations compared to the control. At the 5 mg/L concentration, root growth (dry weight) was reduced by 95% compared with the control value (0.06 g compared with 1.18 g); root length (6 cm) was 26% of the control value (23 cm). Plant height was 18 cm compared with the control value of 28 cm. Rhizome dry weight was affected at concentrations ≥10 mg/L. Although tuber dry weights were greatly reduced compared to controls, the differences were not statistically significant.

Thompson et al. (1997) studied the effects of TNT on tulip polar (*Populus deltoides*) transpiration. Prerooted cuttings were placed in hydroponic solutions containing 0, 1, 3, 5, 10, or 15 mg TNT/L. The solutions were replenished every 1–2 d. At the 5 mg/L concentration, a decrease in growth as indicated by decreased biomass (following chlorotic symptoms and leaf abscission) was evident. Transpiration was affected at 7 d. The 1 mg/L concentration had no effect on growth or transpiration. The authors also reported on the effects of irrigation with a solution of 5 mg/L of TNT on transpiration of larger trees that were planted in sand. The larger trees showed a smaller decrease in transpiration than the laboratory-grown cuttings even though their daily TNT uptake rate was twice that of the cuttings.

Yellow nutsedge was grown in pots containing either Tunica silt or Sharkey clay soils and 80 µg/g (mg/kg, dry weight) of TNT (Pennington 1988a). At 45 d after planting, plant yields of the treated plants from the two soils were higher than yields of controls, but the differences were not statistically significant.

Phaseolus vulgaris (bean), *Triticum aestivum* (wheat), or *Bromus mollis* (blando broom) were grown in pots with TNT at concentrations of 10, 30, or

60 ppm (mg/kg soil, dry weight) (Cataldo et al. 1989). Two soils were tested, Palouse (1.7% organic matter) and Cinebar (7.2% organic matter). Plant height was reduced by >50% in all species in both soils at 60 mg/kg and a reduction of ~25% occurred in the wheat and grass at 30 mg/kg. Marked chlorosis and tip burn were also observed at 60 mg/kg. No phytotoxic symptoms were observed at 10 mg/kg.

During a field survey at the Iowa AAP, a dry lagoon that had not been used for 20 yr was identified. The sediment in the lagoon contained 3030 mg/kg of TNT and was barren of vegetation (Sanocki et al. 1976). Little or no growth of ryegrass, sorghum, and alfalfa took place in soil amended with 1000 or 2000 mg/kg of TNT (Banwart and Hassett 1990).

The root elongation test was conducted with lettuce (*Lactuca sativa*) seeds placed in TNT amended agar (Toussaint et al. 1995). After 96 hr at 5 and 50 mg/L, root tip blackening and negative geotropism occurred, with roots growing away from the agar. The calculated EC_{50} was 2.34 mg/L. Cucumber and radish seedling survival and growth tests were conducted with munitions-contaminated soils from the Joliet AAP (Simini et al. 1995). Toxicity of soil was highly correlated with TNT concentrations although the results may have been influenced by the presence of other contaminants. The LOECs of TNT at two sites within the plant were 7 and 19 mg/kg.

4. Soil Invertebrates. Earthworm survival and growth tests were conducted with munitions-contaminated soils from the Joliet AAP (Simini et al. 1995). Toxicity of soil was highly correlated with TNT concentrations although, as noted earlier, the results may have been influenced by the presence of other contaminants. The LOECs of TNT at two sites within the plant were 7 and 19 mg/kg.

Parmelee et al. (1993) used a soil microcosm to test the effects of TNT on soil fauna communities and trophic structure. TNT was added at 0, 25, 50, 100, or 200 mg/kg to tubes containing soil with natural populations of invertebrates. After 7 d, nematode numbers were increased in the treated soils as a result of greater numbers of hatchlings. Total microarthropod numbers at 200 mg/kg were reduced by 50% compared to controls, but the difference was not statistically significant. The reduction in oribatid mite numbers was statistically significant at this concentration. Only 15% of the applied TNT was extractable after 7 d.

Using several soil types, Phillips et al. (1993) tested the toxicity of TNT to the earthworm *Eisenia foetida* over a 14-d period. Concentrations were 0, 80, 110, 140, 170, and 200 mg/kg in an artificial soil (1.4% organic matter) and 0, 150, 300, 400, and 500 mg/kg in a naturally occurring forest soil (5.9% organic matter). Although survival was 100% for all concentrations in the artificial soil, there was an increasing weight loss with increasing concentrations. Statistical analyses of weight loss data indicated a NOEC of 110 mg/kg and a LOEC of 140 mg/kg. In the forest soil, lethal effects started at 150 mg/kg (7% mortality). The LC_{50} was estimated at 325 mg/kg, and the LOEC was 150 mg/kg.

5. Soil Heterotrophic Processes. The effect of TNT on soil and water microorganisms was studied using solutions. Concentrations greater than 50 mg/L inhibited the growth of most fungi, yeasts, actinomycetes, and gram-positive bacteria (Osmon and Klausmeier 1972; Klausmeier et al. 1973). Most organisms grew when concentrations did not exceed 20 mg/L. Degradation was most rapid when organic nutrients such as yeast extract were added to the medium. Acclimated and unacclimated microbial populations collected from several AAP were not inhibited by 100 mg/L (Jerger et al. 1976). Anaerobic degradation was initially inhibited at 100 mg/L, but not inhibited by 200 mg/L as the acclimation period increased. Jerger et al. (1976) as well as several other authors identified the majority of microorganisms in acclimated cultures as gram-negative bacteria, including *Pseudomonas* spp.

No significant difference was seen in TNT tolerance of actinomycetes (*Streptomyces* spp.) isolated from TNT-contaminated or uncontaminated soils (Pasti-Grigsby et al. 1996). When incubated in nutrient solutions at 25, 50, 75, or 100 mg/L, only a few strains grew above 50 mg/L in yeast malt agar whereas all strains grew at 100 mg/L in a richer agar.

6. Metabolism and Bioaccumulation.

Animals. In mammalian species, TNT is absorbed by inhalation, ingestion, or skin contact. More than 50% of an orally administered dose is absorbed in most test species. Metabolism is extensive and rapid, with 50%–70% excreted in the urine and the remainder in the feces; the majority of the ingested compound is excreted within 24 hr. Distribution to the tissues is less than 1%. Several metabolites including the dinitroaminotoluene, diaminonitrotoluene, hydroxyaminodinitrotoluene, as well as glucuronide-conjugated metabolites, have been recovered from the urine. The urine of some species is bright red in color (ATSDR 1995a).

No laboratory studies on bioaccumulation were located. Rapid metabolism and excretion in mammals (ATSDR 1995a) indicate a low potential for bioaccumulation. Biomonitoring studies have been conducted at Aberdeen Proving Ground, MD (USACHPPM 1994), at the Alabama AAP, Childersburg (Shugart et al. 1990), and at several other AAPs (see Opresko 1995b for review). TNT was not present in tissues of terrestrial wildlife (deer and small mammals) at or above a detection limit of 0.2 mg/kg. Following administration of a subacute dose of TNT of 100 mg/kg to mice, concentrations in liver and muscle were <1.2 mg/kg (Shugart et al. 1990).

Plants. TNT is metabolized by plants grown in TNT-amended soils and solutions as evidenced by the recovery of metabolites such as 2-amino-4,6-dinitrotoluene and 4-amino-2,6-dinitrotoluene from the plant tissues (Palazzo and Leggett 1986; Cataldo et al. 1989; Harvey et al. 1990).

Bioconcentration of TNT by yellow nutsedge was studied in hydroponic solutions containing 5, 10, or 20 mg/L (Palazzo and Leggett 1986). Solutions were replaced after 3 wk. After 42 d, roots, leaves, rhizomes, and tubers were analyzed for TNT and its metabolites, 2-amino-4,6-dinitrotoluene and 4-amino-2,6-dinitrotoluene. TNT was taken up by the plant and, along with its metabolites, was translocated throughout the plants with highest concentrations in the roots. Uptake generally increased with increasing concentrations in the growth medium. At 20 mg/L, concentrations in plant roots were 714, 614, and 2180 mg/kg dry weight for TNT, 2-amino-4,6-dinitrotoluene, and 4-amino-2,6-dinitrotoluene, respectively. At the 20 mg/L concentration, leaves, rhizomes, and tubers contained 13, 95, and 69 mg/kg dry weight of TNT, respectively. Analyses for metabolites in the growth medium were negative.

^{14}C-labeled-TNT added to soil (80 mg/kg) was used to study uptake by yellow nutsedge (Pennington 1988a). Uptake was limited, and the compound did not bioaccumulate. Measurement of plant radioactivity did not allow identification of specific compounds. Lack of bioavailability was attributed to loss from soil by volatilization of degradation products and adsorption of TNT and its metabolites to soil.

E. Terrestrial Criteria and Screening Benchmarks

1. Mammals. A human oral Reference Dose (RfD) of 0.0005 mg/kg/d (USEPA 1993c) was identified by extrapolation (and multiplication by uncertainty factors) from the subchronic study by Levine et al. (1990) in which dogs were administered TNT by capsule. In that study, the endpoint was liver injury.

Several subchronic and chronic studies were available for calculating screening benchmarks for terrestrial mammals. These studies along with identified NOAELs and LOAELs are summarized in Table 12. Testicular atrophy was selected as the endpoint that diminishes wildlife population growth or survival. The study by Dilley et al. (1982) in which Sprague-Dawley rats were administered TNT in the diet at doses of 0, 1.40, 6.97, 34.7 or 160 mg/kg/d for 13 wk was chosen to derive screening benchmarks for representative wildlife because it identifies a higher NOAEL than the study by Levine et al. (1984). Although signs of anemia and some organ weight changes occurred at the two intermediate dose levels, the testicular atrophy observed at the highest dose is considered to be more relevant for population level effects. Therefore, 160 mg/kg/d is considered to be a subchronic LOAEL. The authors state that testicular "lesions did not occur or were far less frequent in other groups"; however, it is not clear whether the dose level of 34.7 mg/kg could be considered the NOAEL for these effects. Therefore the NOAEL was estimated by multiplying the LOAEL by an uncertainty factor of 0.1 and the chronic NOAEL was estimated by multiplying the subchronic NOAEL by 0.1. The final chronic NOAEL is 1.60 mg/kg/d.

Body weights were reported for weeks 0, 1, 2, 4, 8, and 13. For male rats in the high-dose group, average body weight ranged from 0.170 kg at the start of the study to 0.369 kg at 13 weeks. The overall geometric mean for each reported time period was derived from the geometric mean for each reported time interval. This mean value of 0.289 kg was used in the calculations.

Screening benchmarks for oral intake for seven selected wildlife species were derived using the methodology described in the Introduction. An example of the methodology using the meadow vole follows. To calculate the NOAEL for the meadow vole from the NOAEL for the laboratory rat, the following equation was used:

$$\text{Wildlife NOAEL} = \text{test NOAEL} \times \left[\frac{\text{test bw}}{\text{wildlife bw}}\right]^{1/4}$$

where: wildlife bw = body weight of meadow vole = 0.044 kg
test bw = body weight of laboratory rat = 0.289 kg
test NOAEL = experimental dose = 1.6 mg/kg/d

Therefore:

$$\text{Meadow vole NOAEL} = 1.6\ \text{mg/kg/d} \times \left[\frac{0.289\ \text{kg}}{0.044\ \text{kg}}\right]^{1/4} = 2.56\ \text{mg/kg/d}$$

The food factor (f) for meadow voles is 0.114; therefore, the dietary screening benchmark (Cf) is equivalent to

$$\text{Cf} = \frac{\text{Meadow vole NOAEL}}{f} = \frac{2.56}{0.114} = 22.47\ \text{mg/kg food}$$

The water factor (ω) for meadow voles is 0.136; therefore, the drinking water screening benchmark for drinking water (Cw) is equivalent to

$$\text{Cw} = \frac{\text{Meadow vole NOAEL}}{\omega} = \frac{2.36}{0.136} = 18.8\ \text{mg/L water}$$

To calculate a screening benchmark for TNT for a piscivorous species like mink, the following equation is used:

$$\text{Cw} = \frac{\text{NOAEL}_w \times \text{bw}_w}{\text{W} + (\text{F} \times \text{BCF} \times \text{FCM})}$$

The chronic NOAEL for mink is derived from the chronic NOAEL of 1.6 mg/kg for the laboratory rat:

$$\text{Wildlife NOAEL} = \text{test NOAEL} \times \left[\frac{\text{test bw}}{\text{wildlife bw}}\right]^{1/4}$$

where: wildlife bw = body weight of mink = 1.0 kg
test bw = body weight of laboratory rat = 0.289 kg
test NOAEL = experimental dose = 1.6 mg/kg/d

Therefore:

$$\text{Mink NOAEL} = 1.6\ \text{mg/kg/d} \times \left[\frac{0.289\ \text{kg}}{1.0\ \text{kg}}\right]^{1/4} = 1.173\ \text{mg/kg/d}$$

The BCF for TNT can be estimated from the log K_{ow} of 2.03 (see Table 4) by the following equation:

$$\log \text{BCF} = 0.76 \log K_{ow} - 0.23 = 1.3128$$
$$\text{BCF} = 20.55$$

The log K_{ow} of 2.03 is also used to estimate the FCM. For this log K_{ow}, the FCM is 1.0. For mink with a body weight of 1.0 kg, a food consumption rate (F) of 0.137 kg/d, and a water ingestion rate (W) of 0.099 L/d (see Table 3), the overall screening benchmark for water (Cw) is

$$\text{Cw} = \frac{1.173 \times 1.0}{[0.099 + (0.137 \times 20.55 \times 1.0)]} = 0.40\ \text{mg/L}$$

The screening benchmarks for the wildlife species are listed in Table 13. The dietary concentrations in food or water for the wildlife species listed in Table 13 assume exposure through food or water alone and no exposure through other environmental media. If contaminants are present in both food and water, or in other media, the values in Table 13 must be adjusted to sum to the NOAEL.

2. Birds. No suitable subchronic or chronic studies were found for representative avian species.

3. Plants. In the absence of criteria for terrestrial plants, LOEC values from the literature can be used to screen chemicals of potential concern for phytotoxicity (Will and Suter 1995a).

Table 13. TNT screening benchmarks for selected mammalian wildlife species.

Wildlife species	Chronic NOAEL (mg/kg/d)	Screening benchmarks: Diet (mg/kg food)	Water (mg/L)	Piscivorous species (mg/L)[a]
Short-tailed shrew	3.4	5.6	15	—
White-footed mouse	3.0	20	10	—
Meadow vole	2.6	23	19	—
Cottontail rabbit	1.1	5.7	12	—
Mink	1.2	8.6	12	0.40
Red fox	0.8	8.1	9.6	—
Whitetail deer	0.4	14	6.6	—

[a]Water concentration that incorporates dietary intake from both water and food consumption.

Thompson et al. (1997) reported reductions in growth and transpiration of young poplar trees, and Palazzo and Leggett (1986) reported reductions in growth of yellow nutsedge at a solution concentration of 5 mg/L. Therefore, based on these two studies, the screening benchmark for soil water would be 5 mg/L. Pennington (1988a) reported no effects on the growth of yellow nutsedge at a soil concentration of 80 mg/kg (dry weight). However, Cataldo et al. (1989) reported phytotoxic effects at 30 mg/kg but not at 10 mg/kg. Therefore, based on the second study, the LOEC for soil would be 30 mg/kg and the NOEC would be 10 mg/kg (Table 14). Moderate confidence can be placed in benchmarks determined from two studies. However, additional studies are needed in order to determine screening benchmarks with a high degree of confidence.

4. Soil Invertebrates. From a single 7-d soil microcosm study (Parmelee et al. 1993), a NOEC on numbers of soil nematodes and microarthropods of 100 mg/kg and a LOEC of 200 mg/kg TNT were reported. Therefore, using the method of Will and Suter (1995b), the benchmark for other soil invertebrates is 200 mg/kg (Table 14). Confidence in the benchmark is low because of limited data.

A study using the earthworm *Eisenia foetida* (Phillips et al. 1993) determined LOECs for weight loss and death of 140 and 150 mg/kg, respectively. The benchmark of 140 mg/kg (Table 14), based on the sublethal effect of weight loss, is a more appropriate endpoint than lethality. Confidence in the benchmark is low because of limited data.

5. Soil Heterotrophic Processes. In the absence of criteria for soil heterotrophic processes, LOEC data from the literature can be used to screen for chemicals of potential concern (Will and Suter 1995b). Using two different enrichment media, growth of several strains of *Streptomyces* spp. was not inhibited at TNT concentrations of 50 mg/L (Pasti-Grigsby et al. 1996). Klausmeier et al. (1973) found that a variety of soil microorganisms grew when TNT concentrations did not exceed 20 mg/L. Although the microorganisms in this study were isolated from soil at a munitions site, there was no indication that they were acclimated

Table 14. TNT screening benchmarks for terrestrial plants and invertebrates.

Organism/process	Screening benchmark[a]
Plants, solution	5 mg/L
Plants, soil	30 mg/kg
Soil invertebrates (earthworms)	140 mg/kg
Soil invertebrates (other)	200 mg/kg
Soil microbial processes	>20 mg/L

[a]Based on LOEC values.

to TNT. Because the 20 mg/L screening benchmark is based on two studies, confidence in the benchmark is moderate.

III. 1,3,5-Trinitrobenzene

1,3,5-Trinitrobenzene (TNB) is an explosive that is less sensitive to impact than TNT but is more powerful and brisant (Budavari et al. 1996). It has been used to vulcanize natural rubber and as an acid-base indicator in the pH range of 12 to 14 (HSDB 1995b). Chemical and physical properties are presented in Table 15. TNB is formed as a by-product during the manufacture of TNT. It probably arises through the oxidation of the methyl group of TNT to the corresponding acid followed by decarboxylation (Spanggord et al. 1982a). It is present in the final TNT product at concentrations ranging from 0.1% to 0.7% (Wentzel et al. 1979). The photolysis of TNT also results in the formation of TNB (Burlinson et al. 1973; Burlinson 1980), indicating *de novo* synthesis in the environment.

Table 15. Chemical and physical properties of 1,3,5-trinitrobenzene (TNB).

Synonyms	sym-Trinitrobenzene	HSDB 1995b
CAS number	99-35-4	HSDB 1995b
Molecular weight	213.11	Budavari et al. 1996
Physical state	Orthorhombic bipyramidal plates from glacial acetic acid	Budavari et al. 1996
Chemical formula	$C_6H_3N_3O_6$	Budavari et al. 1996
Structure	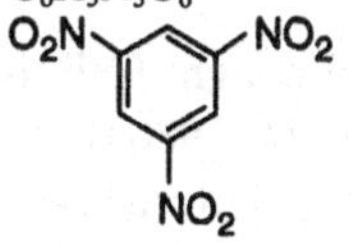	Budavari et al. 1996
Water solubility (20 °C)	0.34 g/L	Spanggord et al. 1980a
Specific gravity (20 °C)	1.76	Budavari et al. 1996
Melting point	122.5 °C	Budavari et al. 1996
Boiling point	315 °C; explodes when heated	HSDB 1995b; Budavari et al. 1996
Vapor pressure (20 °C)	3.2×10^{-6} mm Hg 2.2×10^{-4} torr, 5.1×10^{-6} torr	Spanggord et al. 1980a
Partition coefficients		
Log K_{ow}	1.18	Hansch and Leo 1985
	1.36	Spanggord et al. 1978
Log K_{oc}	1.8–2.7	Spanggord et al. 1980a; ATSDR 1995b; HSDB 1995b
	2.0, 2.25 (estimated)	SRC 1995a
Henry's law constant (20 °C)	1.3 torr M^{-1} (estimated)	Spanggord et al. 1980a
(25 °C)	3.08×10^{-9} atm-m^3/mole (estimated)	HSDB 1995b

Few data on transport and transformation processes in the environment are available. TNB appears to be resistant to photolysis and hydrolysis; microbial reduction of the nitro groups occurs but not ring cleavage. Data were available to develop Tier II aquatic water quality criteria and a SQB. A chronic feeding study was used to develop screening benchmarks for mammalian wildlife. Two subchronic feeding studies, one using the laboratory rat and the other using a wildlife species, were used to compare the sensitivity of laboratory and wildlife species to chemicals. The database on mammalian toxicity testing also included reproductive and developmental studies. No studies were available for calculating screening benchmarks for terrestrial plants, soil invertebrates, or soil microorganisms.

A. Environmental Fate

1. Sources and Occurrences. TNB is released to the environment from munitions production and processing facilities. At a TNT production facility, TNB was detected in 3.8% of samples of condensate wastewater at concentrations of 0.06–0.20 mg/L (Spanggord et al. 1982a). As of 1995, ATSDR reported that TNB had been identified at 14 National Priorities List sites across the U.S. (ATSDR 1995b); these include AAPs as well as LAP plants.

Air. No monitoring data on TNB in air were located. A low vapor pressure of 3.2×10^{-6} mm Hg at 20 °C, estimated by Spanggord et al. (1980a), indicates that TNB is not likely to partition to air.

Surface Water and Groundwater. Although TNB is only slightly soluble in water, 0.34 g/L at 20 °C (Spanggord et al. 1980a), it has been found in surface water and groundwater. Sullivan et al. (1978) measured TNB in Waconda Bay, the receiving water body of the Volunteer AAP, TN. The concentrations in the water ranged from <0.75 μg/L (the limit of detection) to 66 μg/L, and concentrations in the sediment ranged from 73 μg/kg to 300 μg/kg. Within the first mile of the discharge, there was a tendency for concentrations to increase with distance from the discharge, possibly indicating formation of TNB. Concentrations in an onsite stream at the Iowa AAP ranged up to 3.0 μg/L, but concentrations in most of the samples were below the limit of detection of 0.2 μg/L; average concentrations in stream sediment cores ranged up to 5.1 mg/kg (Jerger et al. 1976; Sanocki et al. 1976). The concentration in a single water sample from Boone Creek, which flows through the Louisiana AAP, was 4.8 μg/L (US Army 1987b). Concentrations in a waste ditch and onsite stream at the Joliet AAP ranged up to 16 and 97 μg/L, respectively; concentrations in the respective sediments were <0.1 and up to 3 mg/kg (Jerger et al. 1976).

Detectable concentrations measured in groundwater below leaching pits at the Louisiana AAP ranged from 0.8 to 7720 μg/L (US Army 1987b). Onsite groundwater at the Milan AAP contained concentrations ranging from nondetectable to 976 μg/L (ATSDR 1995b). Maximum concentrations in water from onsite and offsite wells at the Cornhusker AAP were 352 and 114 μg/L, respectively (Monnot et al. 1982; ATSDR 1995b).

Soils. Detectable concentrations of TNB in soil at the Alabama AAP ranged from <0.4 to 3.9 mg/kg (Rosenblatt and Small 1981). Areas sampled included a smokeless powder manufacturing area, magazine area, flashing ground, and an aniline sludge basin. At the Joliet AAP, soil concentrations ranged from less than the limit of detection to 200 mg/kg, with highest concentrations next to an open burning area (Phillips et al. 1994; Simini et al. 1995). The soil concentration in a dry wastewater lagoon at the Iowa AAP was 0.6 mg/kg (Jerger et al. 1976; Sanocki et al. 1976). In a later study at the same site, mean concentrations in a dry lagoon ranged from 0.27 to 0.45 mg/kg, and concentrations in soil taken from an old ordnance burning area ranged from 51 to 62 mg/kg (Jenkins and Grant 1987). At the Cornhusker AAP, shallow and deep soil samples collected at cesspool and leaching pit sites ranged from 13.4 to 1110 mg/kg soil (dry weight) (Monnot et al. 1982). In a survey of TNT and transformation products in a limited number of soil samples collected at AAP, depots, and arsenals, the following air-dried soil concentrations of TNB were found: Nebraska Ordnance Works, 0.12–159 mg/kg; Umatilla Depot, 9.5–63.5 mg/kg; Weldon Springs, 0.3–60.7 mg/kg; Iowa AAP, 53.2–549 mg/kg; Raritan Arsenal, 0.12–3.9 mg/kg; Hawthorne AAP, 3.2–116 mg/kg; Hastings East Park, 2.7 mg/kg (one sample); Milan AAP, 2.5–6.1 mg/kg; and Louisiana AAP, 2.1–3.8 mg/kg (Walsh and Jenkins 1992). TNB was either not present or below the limit of detection at five other installations.

2. Transport and Transformation Processes.

Abiotic Processes. No data on fate in air were located.

Based on a Henry's law constant of 1.3 torr/M, Spanggord et al. (1980a) estimated a volatilization half-life from water of 130 d. Using a group structural estimation method, Syracuse Research Corporation estimated a Henry's law constant of 3.08×10^{-9} atm-m^3/mole at 25 °C (HSDB 1995b). These values suggest that volatilization from water is not a significant transport process. According to Lyman et al. (1982), aromatic nitro compounds are generally resistant to hydrolysis, and Spanggord et al. (1980a) stated that hydrolysis of TNB is not expected to occur under environmental conditions. In a study of the stability of photoproducts of TNT, Burlinson et al. (1973) reported that an aqueous solution of TNB was unchanged after irradiation with a mercury arc lamp for 6 hr.

No data on experimentally derived soil or sediment partition coefficients were located. Based on chemical structure, water solubility, and K_{ow}, K_{oc} of 76–520 have been estimated (Spanggord et al. 1980a; ATSDR 1995b; HSDB 1995b). Syracuse Research Corporation (SRC 1995a) estimated K_{oc} values of 104 and 178 using the regression equations of Lyman et al. (1982) and based on a log K_{ow} of 1.18 (Hansch and Leo 1985) and a water solubility of 340 mg/L at 20 °C (Spanggord et al. 1980a), respectively. These values suggest low to moderate adsorption to soils and suspended sediment in water and moderate to high mobility in soils.

Mitchell et al. (1982) also conducted 3-d bioadsorption studies with TNB and viable or heat-killed cells of *Escherichia coli, Bacillus cereus, Serratia marcescens*, and *Azobacter beijerinckii.* Bioadsorption coefficients ([μg absorbed/gram bacteria]/[μg chemical/mL supernatant]) for viable and heat-killed cells were 6.5 and 7.9, respectively, indicating little bioadsorption.

Biotransformation. Microbial transformation of TNB involves reduction of the nitro groups to form amino groups. This process proceeds through a nitroso intermediate to hydroxylamino compounds and then amines (see Fig. 1). Experiments conducted with cell-free extracts of the bacterium *Veillonella alkalescens* indicate that nitro reduction occurs readily (McCormick et al. 1976), potentially forming triaminobenzene (Wentzel et al. 1979). Although oxidative deamination may take place, the resulting hydroxyl groups would not be ortho to each other and thus ring cleavage would not take place (Wentzel et al. 1979).

Microbial transformation also results in formation of 3,5-dinitroaniline (Mitchell et al. 1982). The metabolite was formed when TNB was added to nutrient-enriched water samples from the Tennessee River taken downstream of the Volunteer AAP. Incubation of 3,5-dinitroaniline under the same conditions as that for TNB resulted in 3,5-diaminonitrobenzene (the principal product), 3-nitro-5-aminoacetanilide, and a nitro, amino-substituted *N*-methylindoline.

Incubation of TNB with *Pseudomonas* sp. isolated from TNT-contaminated soil resulted in the following sequential metabolites: 1,5-dinitroaniline, dinitrobenzene, and 5-nitrobenzene (Boopathy et al. 1994). Nitrobenzene and ammonia accumulated in the medium with no further mineralization. Thus, under aerobic conditions, TNB served as the sole source of nitrogen, but not as a source of carbon.

TNB was resistant to complete mineralization in several other studies. Incubation of TNB at an initial concentration of 100 mg/L for 180 min in a Warburg respirometer inoculated with a phenol-adapted mixed culture of microorganisms obtained from garden soil, compost, river sediment, and a petroleum refinery resulted in little oxygen uptake (Tabek et al. 1964). Tests with ^{14}C-labeled TNB did not result in liberation of $^{14}CO_2$ (Mitchell et al. 1982). Mitchell et al. (1982) conducted microbial screening tests using Tennessee River water collected downstream of the Volunteer AAP. In tests lasting 19 d, concentrations of TNB (10 μg/mL) decreased 4%–6% in filtered river water, 9% in river water with a normal sediment load, and 24% in river water with the sediment enriched threefold. In other studies, TNB did not serve as a sole carbon source for growth of microorganisms.

B. Aquatic Toxicology

1. Acute Effects: Invertebrates and Fish. Acute tests were conducted under static conditions with one invertebrate and four species of fish (Table 16). For

Table 16. Acute toxicity of TNB to aquatic invertebrates and fish.[a]

Test species	Test duration	EC_{50}[b] or LC_{50} (mg/L)	Reference
Daphnia magna (water flea)	48-hr	2.7	Pearson et al. 1979; Liu et al. 1983a
Daphnia magna (water flea)	48-hr	2.98[c]	van der Schalie 1983
Pimephales promelas (fathead minnow)	96-hr	1.1	Pearson et al. 1979; Bailey and Spanggord 1983; Liu et al. 1983a
Pimephales promelas (fathead minnow)	96-hr 10-d[d]	0.49[c] 0.45[c]	van der Schalie 1983
Lepomis macrochirus (bluegill sunfish)	96-hr	0.85[c]	van der Schalie 1983
Oncorhynchus mykiss (rainbow trout)	96-hr 10-d[d] 18-d[d]	0.52[c] 0.52[c] 0.43[c]	van der Schalie 1983
Ictalurus punctatus (channel catfish)	96-hr	0.38[c]	van der Schalie 1983

[a]All tests were conducted under static conditions except where otherwise noted.
[b]EC_{50} values are for *Daphnia magna*.
[c]Measured concentration.
[d]Flow-through test.

both daphnid and fish tests, static acute methods generally followed those recommended by the American Society for Testing and Materials (ASTM 1980). The 48-hr EC_{50} values obtained in two different laboratories for unfed *Daphnia magna* were almost identical (i.e., 2.7 and 2.98 mg/L). Fish were more sensitive to TNB than daphnids, with 96-hr LC_{50} values of 0.38–1.1 mg/L. Channel catfish appeared to be the most sensitive species. Longer-term acute values, in which flow-through conditions were used, were also obtained with fathead minnows and rainbow trout. The 10-d LC_{50} values were very close to 96-hr values.

2. Chronic Effects: Invertebrates and Fish. Chronic tests were conducted under flow-through conditions with one invertebrate and two species of fish (Table 17). Daphnids were tested at measured concentrations of 0 to 2.68 mg/L, fathead minnows were tested at measured concentrations of 0.08 to 0.72 mg/L, and rainbow trout were tested at measured concentrations of 0.08 to 0.71 mg/L. TNB was very toxic to fish, with LOECs for several parameters <1.0 mg/L.

3. Plants. Only one species of algae (*Selenastrum capricornutum*) was tested for toxicity to TNB. Under static conditions, mortality or significantly reduced

Table 17. Chronic toxicity of TNB to aquatic invertebrates and fish.[a]

Species	Stage/age	Parameter measured	Parameter response	Concentration (mg/L)	Reference
Daphnia magna (water flea)	Neonates	Survival and reproduction (21 d)	No effect (NOEC)	0.47	van der Schalie 1983
			Decreased survival	>2.68	
			Decreased reproduction (LOEC)	0.75	
Pimephales promelas (fathead minnow)	Egg/fry	Time to hatch, hatching success, fry survival, overall survival, fry deformities, fry length, fry weight (32 d)	No effect (NOEC)	0.08	van der Schalie 1983
			Decreased survival (LOEC) (fry and overall)	0.12	
Oncorhynchus mykiss (rainbow trout)	Egg/fry	Time to hatch, hatching success, fry survival, overall survival, fry deformities, fry length, fry weight (71 d)	No effect (NOEC)	0.08	van der Schalie 1983
			Decreased hatching success	>0.17	
			Decreased fry survival (LOEC)	0.17	
			Decreased overall survival	0.17	
			Decreased fry length	0.17	
			Decreased fry weight	0.17	

[a]Tests were conducted under flow-through conditions; all concentrations were measured.

growth, measured on d 5 and 14 of the test, occurred at all concentrations tested, 0.10–17.3 mg/L (Bailey 1982b). Concentrations of 0.10–0.17 were algistatic and concentrations of 1.18–17.3 mg/L resulted in mortality. Although a no-effect concentration was not attained in this study, it would be <0.10 mg/L.

4. Metabolism and Bioconcentration. No studies on metabolism or bioconcentration in aquatic organisms were located. Based on a hydrophobic fragment method, Deneer et al. (1987) calculated a log P (log K_{ow}) of 1.18. Liu et al. (1983a) used an estimated log P of 1.36 (Spanggord et al. 1978) to calculate a steady state BCF for organisms with a lipid content of 8%. The calculated BCF of 6.36 indicates little propensity to bioconcentrate.

C. Aquatic Criteria and Screening Benchmarks

1. Aquatic Organisms. Insufficient data were available for calculation of acute and chronic WQC according to USEPA guidelines (Stephan et al. 1985). Data are available for only four of the required eight families: a planktonic crustacean (*D. magna*), the family Salmonidae (*Oncorhynchus mykiss*), a second family in the class Osteichthyes (*Lepomis macrochirus*), and a third family in the phylum Chordata (*Pimephales promelas* or *Ictalurus punctatus*). For three of the families, data were available to calculate acute chronic toxicity ratios: Daphnidae, Salmonidae, and Cyprinidae. Therefore, Tier II or secondary acute and chronic values were calculated according to USEPA guidance for the Great Lakes System (USEPA 1993a). However, it should be noted that the acute and chronic tests with fish were by the same investigator; the toxicity tests with fish were limited to early life stage tests. Acute tests were conducted under static conditions, but TNB is stable in water. The methodology and calculations follow.

The SAV was calculated by dividing the lowest GMAV in the data base by the SAF. GMAVs (geometric means if multiple test results were reported) for the species in Table 16 are *D. magna*, 2.84 mg/L; *P. promelas*, 0.73 mg/L; *L. macrochirus*, 0.85 mg/L; *O. mykiss*, 0.52 mg/L; and *I. punctatus*, 0.38 mg/L. The lowest GMAV was 0.38 mg/L for *I. punctatus*. USEPA (1993a) lists a SAF of 6.5 for use in Tier II calculations when four satisfied data requirements for Tier I calculations are available.

Therefore,

$$\mathrm{SAV} = \frac{0.38}{6.5} = 0.06\ \mathrm{mg/L}$$

The SMC is one-half of the SAV or 0.03 mg/L.

For determination of the SCV, three experimentally determined ACRs were available according to guidelines for Tier I (Table 18). Chronic values are the geometric mean of the LOEC and NOEC for each species, e.g., the geometric mean of 0.47 and 0.75 mg/L = 0.59 mg/L for daphnids. The SACR is the geometric mean of 4.81, 7.58, and 4.46 or 5.46 mg/L. The SCV is the SAV divided by the SACR:

Table 18. Aquatic acute/chronic ratios (ACR) for TNB.

Species	Acute values (mg/L)	Chronic values (mg/L)	Acute/chronic ratios (ACR)
Daphnid	2.84	0.59	4.81
Fathead minnow	0.734	0.098	7.58
Rainbow trout	0.52	0.117	4.46

$$SCV = \frac{SAV}{SACR} = \frac{0.06}{5.46} = 0.011 \text{ mg/L}$$

The acute and chronic screening benchmarks are listed in Table 19.

For aquatic plants, only one species of algae, *Selenastrum capricornutum*, was tested. Because a no-effect level was not reached in this test, the lowest tested concentration is listed in Table 19.

2. *Sediment-Associated Organisms.* The sediment screening benchmark was calculated using the EqP method of Di Toro et al. (1991) and USEPA (1993b): $SQB_{oc} = K_{oc} \times FCV$ (or SCV if FCV cannot be calculated). Because a K_{ow} value is available, a K_{oc} can be calculated and a SQB_{oc} can be determined. Using the log K_{ow} value of 1.36 (Table 15) and the relationship between K_{ow} and K_{oc}, the calculated log K_{oc} is 1.34 and the K_{oc} is 21.7. The SCV as calculated earlier is 0.011 mg/L. Then, the organic carbon-normalized SQB, i.e., mg/kg organic carbon (mg/kg_{oc}), is

$$SQB_{oc} = K_{oc} \times SCV$$
$$SQB_{oc} = 21.7 \text{ L/kg} \times 0.011 \text{ mg/L} = 0.24 \text{ mg/kg}_{oc}$$

Table 19. Water quality criteria/screening benchmarks for TNB.

Criterion	Value (mg/L)
Acute water quality criterion	Insufficient data
Chronic water quality criterion	Insufficient data
Secondary acute value	0.06
Secondary maximum concentration	0.03
Secondary chronic value	0.011
Secondary continuous concentration	0.011
Lowest chronic value, fish	0.12
Lowest chronic value, daphnids	0.75
Lowest plant value	0.10
Sediment quality benchmark (SQB_{oc})	0.24[a]

[a]mg of chemical/kg organic carbon in sediment.

If an organic carbon content in the soil of 1% is assumed, the $SQB_{1\%}$ is calculated as:

$$SQB_{1\%} = 0.01 \times 21.7 \text{ L/kg} \times 0.011 \text{ mg/L} = 0.002 \text{ mg/kg}$$

D. Terrestrial Toxicology

1. Mammals. One chronic and two subchronic studies as well as reproductive and developmental toxicity studies were located; one of the subchronic studies used the laboratory rat and the other a wildlife species. All studies were well conducted and followed USEPA-recommended guidelines. Although chronic studies are preferable for determining long-term effects on populations, the two subchronic toxicity studies are discussed here to compare the sensitivity of laboratory test animals with a wildlife species. These studies were reviewed by Reddy et al. (1997).

In the first subchronic toxicity study, groups of 10 male and 10 female Fischer-344 rats were administered 0, 66.67, 400, or 800 mg TNB/kg diet for 90 d (Reddy et al. 1994a). Calculated doses were 0, 3.91, 22.73, and 44.16 mg/kg/d for males and 0, 4.29, 24.70, and 49.28 mg/kg/d for females. Food intake in the 400- and 800-mg dietary groups was reduced throughout the study and resulted in a significant decrease in absolute body weights. A decrease in testicular weight in males, an increase in relative brain weight in males, and an increase in relative spleen weight of both sexes in the 400- and 800-mg dietary groups were noted. The relative liver weight was increased in the 800-mg dietary groups of both sexes. Histopathological changes included hyaline droplet formation in the kidney, extramedullary hematopoiesis in the spleen, and seminiferous tubular degeneration in the testes (the latter was moderate to severe at 400 and 800 mg TNB/kg diet). Hematology and clinical chemistry studies indicated a decrease in red blood cell count and hematocrit, a decrease in alkaline phosphatase, an increase in reticulocytes, and increased methemoglobin concentration compared to controls of both sexes.

In the second subchronic toxicity study, groups of 10 male and 10 female white-footed mice (*Peromyscus leucopus*) were administered 0, 150, 375, or 750 mg TNB/kg diet for 90 d (Pathology Associates, Inc. 1994). Calculated doses were 0, 23.50, 67.44, and 113.51 mg/kg/d for males and 0, 20.16, 64.81, and 108.25 mg/kg/d for females. The only treatment-related finding in females was an increased relative kidney weight in the group administered 375 mg TNB/kg diet. Biologically significant treatment-related findings were present in males administered 750 mg TNB/kg diet. These changes included an increased relative and absolute spleen weight, erythroid cell hyperplasia in the spleen, a significantly increased number of reticulocytes, an increased number of white blood cells, and seminiferous tubule degeneration in the testes of 3 of 9 males.

In the chronic (2-yr) dietary study, groups of male and female Fischer-344 rats were fed TNB in the diet at concentrations of 0, 5, 60, or 300 ppm (TV Reddy et al. 1996). Based on food consumption data, the authors calculated the intake of TNB at 0, 0.23, 2.68, and 13.31 mg/kg/d (females) and 0, 0.22, 2.64, and 13.44 mg/kg/d (males), respectively. Interim sacrifices occurred at 90 d, 6 mon, and 1 yr. Complete toxicological examinations were performed during these periods. At the interim sacrifices, rats administered 60 ppm showed adverse effects such as increased methemoglobin, erythroid cell hyperplasia, and increased relative organ weights; however, these effects did not persist and were not detected at the end of 2 yr (i.e., the decreased testes weights observed at 90- and 180-d interim sacrifices were reversed). These effects persisted, however, in rats administered the high dosage, 300 ppm.

In the reproductive study, male and female Sprague-Dawley rats were administered a diet containing 30, 150, and 300 mg of TNB/kg of diet (Kinkead et al. 1995). Although relative organ weight changes (spleen, kidney, liver) were observed in females killed after 90 d of exposure and sperm effects (reduced numbers of motile spermatozoa and percent of cells showing a circular motion pattern) were observed in males killed after 28 d of exposure (150 and 300 mg/kg for both sexes), there were no adverse effects on reproductive indices. Likewise, no developmental effects were observed in the offspring of Sprague-Dawley rats administered up to 90 mg/kg/d by oral gavage (Cooper and Caldwell 1995).

2. Birds. No subchronic or chronic studies were located that reported the toxicity of TNB to birds.

3. Plants. No studies were located that reported the toxicity of TNB to terrestrial plants.

4. Soil Invertebrates. No studies were located that reported the toxicity of TNB to terrestrial invertebrates.

5. Soil Heterotrophic Processes. No recent data were located on the effects of TNB on soil microorganisms. Several older studies were conducted with soil microorganisms in solution. Simon and Blackman (1953) found that the EC_{50} for inhibition of growth of the fungus *Trichoderma viride* was 21 mg/L. A survey of the literature by Wentzel et al. (1979) found EC_{50} values for bacteria and fungi of 1–100 mg/L, but these studies were reported in non-English journals and no experimental details were reported in the English abstracts.

Metabolism studies were generally conducted at low concentrations, and conclusions regarding toxicity cannot be made. Furthermore, the cultures were not only adapted to the test compound but a carbon source and nutrients were added. Mitchell et al. (1982) conducted several metabolism studies with Tennessee river water and sediments collected below the Volunteer AAP. At an initial concentration of 10.5 μg/L, TNB underwent primary biodegradation without a lag period. When cultures were enriched with organisms adapted to TNB, concentrations up to 53 μg/L underwent primary biodegradation in 144 h. In another study, an aerobic bacterial consortium obtained from an anoxic soil slurry reactor used to treat TNT-contaminated soil grew in an enriched medium containing 50 mg/L of TNB (Boopathy et al. 1994).

6. Metabolism and Bioaccumulation.

Animals. TNB is absorbed by mammalian species, as evidenced by the detection of its metabolites in biological fluids. Bel et al. (1994) detected 3,5-dinitroaniline (urine), 3-amino-5-nitroaniline (urine, feces, and blood), and 1,3,5-triaminobenzene (urine and feces) in rats fed TNB in the diet. In an *in vitro* study, G. Reddy et al. (1996) detected 3,5-dinitroaniline and 3-amino-5-nitroaniline in rat liver microsomal preparation; TNB added to the system was metabolized within 5 min.

No laboratory studies on bioaccumulation were located. Biomonitoring studies have been conducted at Aberdeen Proving Ground, MD (USACHPPM 1994), and at several other AAPs (see Opresko 1995b for review). TNB was not present in tissues of terrestrial wildlife (deer and small mammals) at or above a detection limit of 0.2 mg/kg.

Plants. No data were located on metabolism and bioaccumulation by terrestrial plants.

E. Terrestrial Criteria and Screening Benchmarks

1. Mammals. A human oral (RfD) of 0.03 mg/kg/d (USEPA 1997) was identified by extrapolation (and multiplication by a total uncertainty factor of 100) from the chronic study by TV Reddy et al. (1996) in which Fischer-344 rats were administered TNB in the diet. The endpoints were methemoglobinemia and spleen-erythroid cell hyperplasia.

Three studies are available for calculating TNB screening benchmarks for wildlife; the first is a chronic study using Fischer-344 laboratory rats. Of the subchronic studies, one used the Fischer-344 rat and the other used a wildlife species, the white-footed mouse. In both subchronic studies the endpoint selected as being relevant to population-level effects was degeneration of the seminiferous tubules of the testes. Because the relative sensitivity of laboratory-reared test species and wildlife species is of interest, screening benchmarks were determined using the data from each study. The studies, with identified NOAELs and LOAELs, are summarized in Table 20.

In the subchronic study with Fischer-344 rats (Reddy et al. 1994a), the dosages were 0, 66.67, 400, and 800 mg/kg diet (laboratory chow) and the calculated average TNB doses for males were 0, 3.91, 22.73, and 44.16 mg/kg/d, respectively. Moderate to severe seminiferous tubular degeneration was seen in males receiving 22.73 or 44.16 mg TNB/kg bw/d. No adverse effects occurred at 3.91 mg/kg/d. The final subchronic NOAEL was 3.91 mg/kg/d and the final chronic NOAEL was 0.4 mg/kg/d (the subchronic NOAEL was multiplied by an uncertainty factor of 0.1 to derive the chronic NOAEL). The mean body

Table 20. TNB toxicity data for mammalian species.

Species	Route	Exposure period	NOAEL (mg/kg/d)	LOAEL (mg/kg/d)	Effect/endpoint	Reference
Rat	Diet	13 wk	3.91	22.73	Testicular effects	Reddy et al. 1994a
	Diet	2 yr	2.64	13.44	Testicular effects	TV Reddy et al. 1996
White-footed mouse	Diet	13 wk	67.44	113.51	Testicular effects	Pathology Associates, Inc. 1994

weight (geometric mean of weekly means for the 13-wk test period) for the 3.91 mg/kg male test group was calculated to be 274 g.

In the subchronic study with the white-footed mouse (Pathological Associates Inc. 1994), the dosages were 0, 150, 375, and 750 mg/kg diet and the calculated average daily TNB doses for males in each group were 0, 23.50, 67.44, and 113.51 mg/kg bw, respectively. Seminiferous tubular degeneration was seen in 3 of 9 males receiving 113.51 mg TNB/kg bw. No adverse testicular effects occurred in mice receiving 23.50 or 67.44 mg/kg bw. The final subchronic NOAEL was 67.44 mg/kg/d and the final chronic NOAEL was 6.74 mg/kg/d (the subchronic NOAEL was multiplied by an uncertainty factor of 0.1 to derive the chronic NOAEL). The authors reported that the final average bw of male mice receiving 67.44 mg TNB/kg bw was 18.5 g.

In the chronic study (TV Reddy et al. 1996), male and female rats in the high-dose group showed decreased body weights associated with decreased food consumption, changes in relative organ weights, and adverse hematological findings; however, the life span was not affected and testicular effects, evident at the shorter time periods, were reversed except in the high-dose group. These results, coupled with generally negative findings in reproductive and developmental toxicity studies (Kinkead et al. 1995; Cooper and Caldwell 1995), indicate that the highest dose tested, 13.44 mg/kg/d in males, is a LOAEL for ecologically relevant endpoints and the middose is a NOAEL. Because this was a chronic study, a subchronic to chronic uncertainty factor was not applied, and the final chronic NOAEL was 2.64 mg/kg/d. The average body weight of males during the study was 350 g.

Screening benchmarks for food and water intake for selected wildlife species were derived from each study using the methodology described in the Introduction and illustrated in the previous section on TNT. The log K_{ow} value of 1.36 was used to calculate the BCF for mink. Screening benchmarks derived from the subchronic study with the laboratory rat and white-footed mouse are listed in Tables 21 and 22, respectively; screening benchmarks derived from the chronic study with the laboratory rat are listed in Table 23.

Although the test species chronic NOAELs differ somewhat between the chronic study with the rat and the subchronic study with the white-footed mouse (2.64 and 6.74 mg/kg/d, respectively), the differences in test species body size used in the calculations result in very similar screening benchmarks (Tables 22 and 23). The data also show that the laboratory rat is more sensitive to the effects of TNB than the white-footed mouse. Because toxicity data for most chemicals involve laboratory species, the extrapolated values for wildlife, based on the laboratory rat data, can be used to compare toxicities among munitions chemicals. On the other hand, the values based on the white-footed mouse may be more realistic for species adapted to the stress and contaminants in their natural environment. The white-footed mouse has not been colonized or inbred for as many generations as the albino mouse and so may be similar to its wild counterparts. Confidence in the benchmarks is high because two different species were tested, one study was a chronic study, and reproductive and developmental studies were also available.

Table 21. TNB screening benchmarks for selected mammalian wildlife species based on subchronic study using the laboratory rat.[a,b]

Wildlife species	Chronic NOAEL (mg/kg/d)	Screening benchmarks		
		Diet (mg/kg food)	Water (mg/L)	Piscivorous species (mg/L)[c]
Shorttail shrew	0.8	1.4	3.8	—
White-footed mouse	0.8	4.9	2.5	—
Meadow vole	0.6	5.5	4.7	—
Cottontail rabbit	0.3	1.4	2.9	—
Mink	0.3	2.1	2.9	0.3
Red fox	0.2	2.0	2.4	—
Whitetail deer	0.1	3.4	1.6	—

[a]Data based on the Fischer-344 rat.
[b]Reddy et al. (1994a).
[c]Water concentration that incorporates intake from both water and food consumption.

2. *Birds.* No suitable subchronic or chronic studies were found for representative avian species.

2. *Plants.* No data were located for calculating a screening benchmark for plants.

3. *Soil Invertebrates.* No data were located for calculating a screening benchmark for soil invertebrates.

Table 22. TNB screening benchmarks for selected mammalian wildlife species based on subchronic study using mammalian wildlife.[a,b]

Wildlife species	Chronic NOAEL (mg/kg/d)	Screening benchmarks		
		Diet (mg/kg food)	Water (mg/L)	Piscivorous species (mg/L)[c]
Shorttail shrew	7.1	12	32	—
White-footed mouse	6.7	38	22	—
Meadow vole	5.4	48	40	—
Cottontail rabbit	2.4	12	25	—
Mink	2.5	18	25	2.6
Red fox	1.7	17	20	—
Whitetail deer	0.9	29	14	—

[a] Data based on the white-footed mouse (*Peromyscus leucopus*).
[b]Pathology Associates, Inc. (1994).
[c]Water concentration that incorporates intake from both water and food consumption.

Table 23. TNB screening benchmarks for selected mammalian wildlife species based on a chronic study with the laboratory rat.[a,b]

Wildlife species	Chronic NOAEL (mg/kg/d)	Screening benchmarks		
		Diet (mg/kg/food)	Water (mg/L)	Piscivorous species (mg/L)[c]
Shorttail shrew	5.8	9.7	26	—
White-footed mouse	5.3	34	18	—
Meadow vole	4.4	39	33	—
Cottontail rabbit	1.9	9.8	20	—
Mink	2.0	15	21	2.1
Red fox	1.4	14	17	—
Whitetail deer	0.7	24	11	—

[a]Data based on the Fischer-344 rat.
[b]TV Reddy et al. (1996).
[c]Water concentration that incorporates intake from both water and food consumption.

5. Soil Heterotrophic Processes. No useful data were located for calculating a soil microbial screening benchmark.

IV. 1,3-Dinitrobenzene

1,3-Dinitrobenzene (DNB) has been used as an explosive and in the manufacture of explosives (Fedoroff et al. 1962). DNB is not a military-unique compound; it is manufactured and used as a chemical intermediate in the synthesis of *m*-nitroaniline and *m*-phenylenediamine, which in turn are chemical intermediates for various products including azo dyes (Benya and Cornish 1994). DNB is manufactured in a two-stage process by the nitration of benzene followed by the nitration of nitrobenzene using a hot mixture of nitric and surfuric acids (HSDB 1995c).

DNB is formed as a by-product during production of TNT, either from the nitration of benzene (an impurity in toluene) or through oxidation and decarboxylation of 2,4-dinitrotoluene (Spanggord et al. 1982a). It is an impurity in the final product and is found in waste discharges at AAPs. Of the total possible isomers of DNB present in TNT, approximately 93% is the 1,3-isomer (Chandler et al. 1972). DNB can also be formed in the environment by photolysis of 2,4-dinitrotoluene, another by-product released into the environment from the manufacture of TNT (Kitchens et al. 1978).

Data on concentrations and fate in the environment, aquatic toxicity (fish, invertebrates, and algae), and terrestrial toxicity (laboratory animals) were located. These data were used to develop Tier II WQC for aquatic organisms and screening benchmarks for mammalian wildlife. Chemical and physical properties are listed in Table 24.

Table 24. Chemical and physical properties of 1,3-dinitrobenzene (DNB).

Synonyms	*m*-Dinitrobenzene 1,3-Dinitrobenzene 1,3-DNB	HSDB 1995c
CAS number	99-65-0	HSDB 1995c
Molecular weight	168.11	Budavari et al. 1996
Physical state	Yellowish crystals	Budavari et al. 1996
Chemical formula	$C_6H_4N_2O_4$	Budavari et al. 1996
Structure	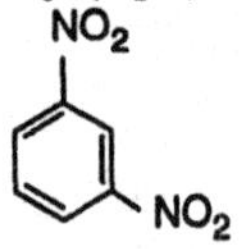	Budavari et al. 1996
Water solubility (20 °C)	500 mg/L	Budavari et al. 1996
Specific gravity	1.575	HSDB 1995c
Melting point	89–90 °C	Budavari et al. 1996
Boiling point	300–303 °C	Budavari et al. 1996
Vapor pressure (20 °C)	0.0039 mm Hg (extrapolated)	USEPA 1985
Partition coefficients		
Log K_{ow}	1.49	Hansch and Leo 1979
	1.62	Liu et al. 1983a
Log K_{oc}	1.8	Spanggord et al. 1980a
	2.3	US Army 1987a
Henry's law constant (20 °C)	2.3×10^{-6} atm-m^3/mol	HSDB 1995c

A. Environmental Fate

1. Sources and Occurrences. DNB is released to the environment from munitions production and processing facilities and through its use as an intermediate in the synthesis of other chemicals. As of 1995, ATSDR reported that DNB had been identified at 12 National Priorities List sites across the US (ATSDR 1995b).

Air. No data on the occurrence or fate of DNB in air were located.

Surface Water and Groundwater. In a 1-yr monitoring study at the Volunteer AAP, DNB was identified in 97.5% of samples of condensate wastewater (range, 0.20–8.5 mg/L) discharged to the environment. The average concentration in the samples was 2.0 mg/L and the 90th percentile concentration was 4.0 mg/L (Spanggord et al. 1978; Liu et al. 1983a).

DNB is slightly soluble in water, 0.5 g/L at 20 °C (Budavari et al. 1996). Only nondetectable to low concentrations occur in surface water and groundwater at AAPs (Walsh 1990). Sullivan et al. (1978) measured DNB in Waconda Bay, the receiving water body of the Volunteer AAP, TN. Concentrations in the water ranged from below the limit of detection (1 μg/L) to 6 μg/L; concentrations in the sediment ranged from <6.3 to 14 μg/kg. Detectable concentrations in groundwater at the Louisiana AAP ranged from 0.74 to 210 μg/L; the compound was not detected in soil and sediment samples at this site (US Army 1989).

Soils. In a survey of 16 AAPs, arsenals, and depots, DNB was found at 6 sites (15 samples); concentrations ranged from 0.06 to 45.2 mg/kg (Walsh and Jenkins 1992). Highest concentrations were found at the Iowa AAP (45.2 mg/kg [1 sample]), the Umatilla Depot (0.3–29.8 mg/kg), and the Hawthorne AAP (0.2–15 mg/kg).

2. Transport and Transformation Processes.

Abiotic Processes. No data on the fate of DNB in air were located.

Photolysis is not expected to be a significant degradation mechanism. The concentration of DNB in synthetic condensate wastewater exposed to light from a 1200-W lamp (exposure time not given) did not change (Liu et al. 1983a). Solutions of DNB exposed to visible light in a photodegradation reactor or exposed to sunlight for 17 d were stable (Spanggord et al. 1978). However, Simmons and Zepp (1986) calculated a half-life of 23 d for DNB in near-surface freshwater at 40° latitude. Additional experiments conducted by Spanggord et al. (1978) in full sunlight and shade indicated that loss of DNB from open containers results from volatilization; volatilization rate loss was determined to be 1.88×10^{-2} ppm/hr. Volatilization half-lives have been estimated at 73 and 550 d (Spanggord et al. 1980a). Smith et al. (1981) measured the volatilization rate of DNB in the laboratory at room temperature. Solutions of various depths were subjected to mechanical stirring and air flow. A range of volatilization rate constants of 0.001–0.0031/hr indicated a half-life of 9–29 d. A calculated Henry's law constant of 2.3×10^{-6} atm-m^3/mol at 20°C (HSDB 1995c) indicates slow volatilization from water. Aromatic nitro compounds are generally resistant to hydrolysis (Lyman et al. 1982).

Few experimental data regarding adsorption to soil or sediment were located. Using the clay mineral kaolinite (0.06% organic carbon), Haderlein and Schwarzenbach (1993) measured an adsorption coefficient (K_d or K_p) of 1800 L/kg. This value would indicate a moderate potential to adsorb to soil. Calculated K_{oc} values of 64 (Spanggord et al. 1980a), based on water solubility, and of 214 (US Army 1987a), indicate low to moderate adsorption on sediment. Based on mathematical derivations from the K_{ow} and the water solubility, USEPA (1985) predicted that most of the compound will be found in the water column and less than 1% of the compound in sediments.

Biotransformation. Microbial degradation proceeds by reduction of the nitro groups to form amino groups. In the presence of enzymes from the bacterium *Veillonella alkalescens*, the pathway proceeded through a nitroso intermediate to hydroxylamino compounds. Further reduction yielded amino compounds (McCormick et al. 1976).

Incubation of DNB in water samples taken from the Tennessee River downstream from the Volunteer AAP resulted in mineralization to CO_2 and indicated that the microorganisms present could use DNB as a sole carbon source (Mitchell and Dennis 1982; Mitchell et al. 1982). Mineralization did not occur in sterile water or in surface water from several sites in Maryland, indicating the importance of the source of the water. In Tennessee River water, a concentration of 5 μg/L was mineralized in 15 d following a 10-d lag period; under these conditions, a half-life of approximately 1 d was estimated following the lag period. When added to enrichment cultures in the laboratory, the half-life was 9.7 d.

Microbial degradation in sewage effluent occurred under both aerobic and anaerobic conditions (Hallas and Alexander 1983). Under both conditions, significant amounts of nitroaniline were formed. Under aerobic conditions, cultures of the yeast *Candida pulcherrima* (isolated from soil contaminated with DNB) produced *m*-nitrophenol, *m*-aminophenol, resorcinol, and fumaric acid, as well as CO_2, the latter indicating complete degradation (Dey and Godbole 1986).

Mitchell et al. (1982) conducted 3-d bioadsorption studies with DNB and viable or heat-killed cells of *Escherichia coli, Bacillus cereus, Serratia marcescens*, and *Azobacter beijerinckii.* Bioadsorption coefficients ([μg absorbed/g bacteria]/[μg chemical/mL supernatant]) for viable and heat-killed cells were 4.3 and 13.1, respectively, indicating little bioadsorption.

B. Aquatic Toxicology

1. Acute Effects: Invertebrates and Fish. Acute toxicity tests were conducted under static conditions with one species of invertebrate and four species of fish (Table 25). For both daphnid and fish tests, static acute methods generally followed those recommended by the American Society for Testing and Materials (ASTM 1980). As indicated by EC_{50} values for immobilization of 27.4 and 49.6 mg/L, DNB is not highly acutely toxic to *Daphnia magna*. In another study, however, a concentration of 43 mg/L resulted in 50% mortality. When *D. magna* were fed during testing, toxicity was reduced as indicated by an EC_{50} for immobilization of 72 mg/L (Deneer et al. 1989). Various species of fish showed different sensitivities. The most sensitive species appeared to be *Lepomis macrochirus* (96-hr LC_{50}, 1.4 mg/L) and *Oncorhynchus mykiss* (96-hr LC_{50}, 1.7 mg/L).

2. Chronic Effects: Invertebrates and Fish. Chronic tests were conducted with one species of invertebrate and one species of fish (Table 26). Sixteen- and 21-d tests conducted under static-renewal conditions with *D. magna* showed a no-effect concentration of 0.55 mg/L, a significant effect on reproduction at 3.2 mg/L, and an effect on growth at approximately 1 mg/L. Thirty-day tests conducted under flow-through conditions with young *O. mykiss* showed a no-effect

Table 25. Acute toxicity of DNB to aquatic invertebrates and fish.[a]

Test species	Test duration	EC_{50}[b] or LC_{50} (mg/L)	Reference
Daphnia magna (water flea)	48-hr	49.6 (immobilization)	Liu et al. 1983a
Daphnia magna (water flea)	48-hr	27.4[c] (immobilization)	van der Schalie 1983
Daphnia magna (water flea)	48-hr	43[c] (mortality)	Hermens et al. 1984
Ictalurus punctatus (channel catfish)	96-hr	8.1[c]	van der Schalie 1983
Oncorhynchus mykiss (rainbow trout)	96-hr	1.7[c]	van der Schalie1983
Lepomis macrochirus (bluegill sunfish)	96-hr	1.4[c]	van der Schalie 1983
Pimephales promelas (fathead minnow)	96-hr	12.7	Curtis and Ward 1981
Pimephales promelas (fathead minnow)	96-hr	7.0	Bailey and Spanggord 1983; Liu et al. 1983a
Pimephales promelas (fathead minnow)	96-hr	16.8[c]	van der Schalie 1983

[a]All tests were conducted under static conditions.
[b]EC_{50} are for immobilization of *Daphnia magna*.
[c]Measured concentration.

concentration of 0.16 mg/L. When tests were conducted for 69 d beginning with the egg stage, the no-effect concentration was 0.50 mg/L and the lowest-effect concentration for survival and growth was 0.97 mg/L.

3. Plants. Tests lasting from 24 hr to 8 d were conducted with four species of algae (Table 27). In the longer-term tests, all the studies indicated effects on population growth at <1 mg/L (0.17–0.97 mg/L). Details of some of the studies are discussed here. The green alga *Selenastrum capricornutum* was tested for toxicity to DNB at measured concentrations of 0.26, 0.97, 1.58, 10.7, 14.3, and 85.6 mg/L under static conditions (van der Schalie 1983). Significantly reduced growth, measured on d 5 of the test, occurred at all concentrations above 0.26 mg/L. Concentrations between 0.97 and 14.3 mg/L were considered algistatic, as transfer to clean media resulted in renewed growth. Deneer et al. (1989) tested DNB toxicity to the green alga *Chlorella pyrenoidosa* under static-renewal conditions and found an EC_{50} value for maximum yield of 0.24 mg/L.

Fitzgerald et al. (1952) exposed cultures of the blue-green alga *Microcystis aeruginosa*, grown in 125-mL flasks at 22 °C, to DNB for 24 hrs. A concentra-

Table 26. Chronic toxicity of DNB to aquatic invertebrates and fish.

Species	Stage/age	Parameter measured	Parameter response	Effect concentration (mg/L)	Reference
Daphnia magna (water flea)	<24-hr old	Reproduction[a]	16-d EC_{50}	3.2	Deneer et al. 1988
		Growth	16-d NOEC	0.55	
		Growth	16-d EC_{10}	1.2	
Daphnia magna (water flea)	<24-hr old	Immobilization[a]	21-d LOEC	2.0	Deneer et al. 1989
		Reproduction	21-d LOEC	3.2	
		Growth	21-d LOEC	0.99	
Oncorhynchus mykiss (rainbow trout)	Young	Lethality	30-d LC_0[b]	0.16	van der Schalie 1983
		Lethality	30-d LC_{50}	0.37	
Oncorhynchus mykiss (rainbow trout)	Egg/fry	Time to hatch, hatching success, time to swim-up, fry survival, overall survival, fry deformities, fry length, fry weight (69 d)[b]	No effect	0.50	van der Schalie 1983
			Decreased survival	0.97	
			Decreased fry length	0.97	
			Decreased fry weight	0.97	

[a]Static-renewal tests.
[b]Tests were conducted under flow-through conditions; concentrations were measured.

Table 27. Toxicity of DNB to algae.

Test species	Test duration	Effect	Concentration (mg/L)	Reference
Anacystis aeruginosa (blue-green alga)	24-hr	LC_{50}	5.0	Fitzgerald et al. 1952
Anacystis aeruginosa (blue-green alga)	8-d	"Toxicity threshold" (population growth)	0.17	Bringmann and Kuhn 1978
Chlorella pyrenoidosa (green alga)	96-hr	EC_{50} (50% reduction in maximum yield)	0.24	Deneer et al. 1989
Selenastrum capricornutum (green alga)	5-d	No effect on growth	0.26	van der Schalie 1983
		Reduced growth	0.97	
		Algistatic	0.97-14.3	
Scendesmus quadicauda (green alga)	7-d	"Toxicity threshold" (cell multiplication inhibition)	0.70	Bringmann and Kuhn 1980

tion of 5 mg/L resulted in 50% mortality. The 8-d "toxicity threshold" (the concentration at which an inhibitory effect begins; ≥3% below controls) for population growth of *M. aeruginosa* was 0.17 mg/L (Bringmann and Kuhn 1978). When tested in the cell multiplication inhibition test, the 7-d "toxicity threshold" for the green alga *Scendesmus quadicauda* was 0.7 mg/L (Bringmann and Kuhn 1980).

4. Metabolism and Bioconcentration. No data on metabolism by aquatic organisms were located. Log K_{ow} values of 1.49 (Hansch and Leo 1979), 1.62 (Liu et al. 1983a), and 1.52 (Deneer et al. 1989) have been calculated. Using a log K_{ow} value of 1.62, calculated according to the method of Leo et al. (1971), and the BCF equation of Veith et al. (1979), Liu et al. (1983a) computed a BCF for fish of 10.03. This value suggests a low potential for bioconcentration. Deneer et al. (1987) exposed guppies (*Poecilia reticulata*) to DNB for 3 d and measured a BCF of 74 on the basis of fat weight.

C. Aquatic Criteria and Screening Benchmarks

1. Aquatic Organisms. Data that met USEPA guidelines were available for four of the required eight families: a planktonic crustacean (*D. magna*), the family Salmonidae (*O. mykiss*), a second family in the class Osterichthyes (*Lepomis macrochirus*), and a third family in the phylum Chordata (*Pimephales promelas* or *Ictalurus punctatus*). For two of the families, data were available to calculate acute/chronic toxicity rations, Daphnidae and Salmonidae. Therefore, Tier II or secondary acute and chronic values were calculated according to USEPA guidance for the Great Lakes System (USEPA 1993a).

The SAV was calculated by dividing the lowest GMAV in the data base by the SAF. The lowest GMAV was 1.4 mg/L for *L. macrochirus*. USEPA (1993a) lists a SAF of 6.5 for use in Tier II calculations when four satisfied data requirements for Tier I calculations are available.

Therefore,

$$\text{SAV} = \frac{1.4}{6.5} = 0.215 \text{ mg/L}$$

The SMC is one-half of the SAV or 0.108 mg/L.

For the SACR, data for two of the three required families were available to calculate ACRs according to guidelines for Tier I. Therefore, one assumed ACR of 18 was added. Most of the acute and chronic studies were performed by the same investigator, and the chronic studies were performed under flow-through conditions. The SACR is the geometric mean of the three ACRs. For *D. magna*, the acute value was set equal to the geometric mean of the two EC_{50} values for immobilization, the typical endpoint in studies with species of Daphnidae. For the chronic value, the endpoints of growth (16-d NOEC and LOEC for *D. magna*) and survival (69-d NOEC and LOEC for *Oncorhynchus mykiss*) were chosen. Acute and chronic values and the ratios are listed in Table 28.

Table 28. Aquatic acute/chronic ratios (ACR) for DNB.

Species	Acute values (mg/L)	Chronic values (mg/L)	Acute/chronic ratios (ACR)
Daphnid	36.87	0.81	45.38
Rainbow trout	1.7	0.696	2.44
Default ACR[a]			18

[a]A default ACR of 18 is used when data from fewer than three acute and chronic tests are available (USEPA 1993a).

The SACR is the geometric mean of 45.38, 2.44, and 18 or 12.58. The SCV is the SAV divided by the SACR:

$$SCV = \frac{SAV}{SACR} = \frac{0.215}{12.58} = 0.017 \text{ mg/L}$$

Aquatic screening benchmarks are listed in Table 29.

The few chronic studies and the range of the SACR make the SCV, at best, an estimate. Another acute test with *O. mykiss* and additional chronic studies with other species are needed to improve confidence in the SCV screening benchmark. van der Schalie (1983) noted that the 30-d LC_{50} for adult *O. mykiss* (0.37 mg/L) was lower than the NOEC for eggs and fry (0.50 mg/L).

For algal species, a test of at least 96-hr duration may be used; concentrations of test material should be measured and the endpoint should be biologically significant (Stephan et al. 1985). The study by van der Schalie (1983) met these conditions; however, the study by Deneer et al. (1989) resulted in a lower LOEC, 0.24 mg/L, and this value was chosen as the final chronic value for plants.

Table 29. Water quality criteria/screening benchmarks for DNB.

Criterion	Value (mg/L)
Acute water quality criterion	Insufficient data
Chronic water quality criterion	Insufficient data
Secondary acute value	0.22
Secondary maximum concentration	0.11
Secondary chronic value	0.02
Secondary continuous concentration	0.02
Lowest chronic value, fish	0.37
Lowest chronic value, daphnids	0.99
Final plant value	0.24
Sediment quality benchmark	0.67[a]

[a]mg DNB/kg organic carbon in sediment.

2. Sediment-Associated Organisms. Because the only experimental adsorption value for DNB was based on a clay mineral (Haderlein and Schwarzenbach 1993) instead of soils or sediments with organic matter content between 1% and 8% as recommended by USEPA (1979), a SQB based on the K_p value was not calculated.

Because K_{ow} values are available, a K_{oc} can be calculated and a SQB_{oc} can be determined. Using the most recently calculated log K_{ow} value of 1.62 (Liu et al. 1983a) and the relationship between K_{ow} and K_{oc}, the log K_{oc} is 1.593 and the K_{oc} is 39.15. Then, the organic carbon-normalized SQB, i.e., mg/kg organic carbon (mg/kg_{oc}) is

$$SQB_{oc} = K_{oc} \times SCV$$
$$SQB_{oc} = 39.15\ L/kg \times 0.017\ mg/L = 0.67\ mg/kg_{oc}$$

If an organic carbon content in the soil of 1% is assumed, the $SQB_{1\%}$ is calculated as

$$SQB_{1\%} = 0.01 \times 39.15\ L/kg \times 0.017\ mg/L = 0.007\ mg/kg$$

D. Terrestrial Toxicology

1. Mammals. No subchronic or chronic studies on the toxicity of DNB using mammalian wildlife species were located. Three subchronic studies on the toxicity of DNB utilizing laboratory animals were located. In the first study, Linder et al. (1986) administered DNB to groups of 12 male Sprague-Dawley rats by gavage, 5 d/wk, for 12 wk at doses of 0. 0.75, 1.5, 3.0, or 6.0 mg/kg/d (0, 0.54, 1.1, 2.1, or 4.3 mg/kg/d when adjusted for a 7-d/wk exposure). Rats dosed with 4.3 mg/kg/d showed decreased body weight gains and slight ataxia by wk 8; two animals in this group died during wk 10. Doses of 2.1 and 4.3 mg/kg/d resulted in increases in spleen weight accompanied by hemosiderosis, diminished sperm production, decreased weights of the testes and epididymides, seminiferous tubular atrophy, and infertility when mated to untreated females. Although sperm production was also decreased in males treated with 1.1 mg/kg/d, the males were fertile and produced a number of litters similar to that of controls. No effects were observed on sperm production or mating parameters at a dose of 0.54 mg/kg/d.

In the second study, groups of 15 male and 15 female Fischer-344 rats were administered 0, 1, 6, or 30 mg DNB/kg diet for 90 d (5 animals per group were killed at 45 d for hematology and clinical chemistry analyses) (Reddy et al. 1995). Calculated doses were 0.06, 0.35, and 1.73 mg/kg/d for males and 0.07, 0.39, and 1.93 mg/kg/d for females. Food and water consumption were not altered in any dose group, and terminal body weights were similar among groups. Relative organ weights were significantly changed for the spleen (increased in the high-dose group for both sexes) and testes (decreased in the high-dose group). A significant decrease in red blood cell count, hemoglobin, and hematocrit and a significant increase in reticulocyte levels were observed in both sexes in the 30 mg DNB/kg diet dose group. Methemoglobin levels were increased in both sexes receiving the 6 and 30 mg DNB/kg diet. The following histological changes were observed in the 30 mg DNB/kg diet group: cytoplasmic droplets in the kidneys (both sexes), erythroid cell hyperplasia of the spleen (both sexes), and seminiferous tubular degeneration in the testes (males).

In the third study, groups of 20 male and 20 female Charles River rats were administered DNB in drinking water at concentrations of 3, 8, or 20 mg/L for 16 wk (Cody et al. 1981). For males, dose intakes of 0, 0.40, 1.13, or 2.64 mg/kg/d were reported, and, for females, dose intakes of 0, 0.48, 1.32, or 3.10 mg/kg/d were reported. Females in the highest dose group had reduced rates of growth; at the end of 16 wk their mean body weight was significantly lower than that of the control females. Growth was normal in the other groups. Moderate reductions in hematocrit and hemoglobin values in males, observed in the high-dose group after 5 and 10 wk, were similar to those of controls at 90 d. In the groups administered 8 and 20 mg/L, spleens of both sexes were significantly increased in weight, and in the males administered 8 mg/L, the testes were significantly reduced in weight. Histopathological examinations revealed the presence of hemosiderin in the spleen, especially in the group receiving 20 mg/L in the drinking water. Testes were normal in appearance, but there was evidence of a slight to moderate decrease in spermatogenesis in the group receiving 20 mg/L. There was some evidence of increased physical activity of males administered DNB, but the cause and significance of this effect are unknown.

2. Birds. No subchronic or chronic studies on the toxicity of DNB to birds were located. The compound is of low acute toxicity to birds, with LD_{50} values >100 mg/kg for starlings and 42 mg/kg for red-winged blackbirds (Schafer 1972).

3. Plants. No data on the toxicity of DNB to terrestrial plants were located. Mature hydroponically grown soybeans (*Glycine max*) showed no phytotoxic symptoms during 4 d in a solution of 34 µg/L of DNB (McFarlane et al. 1987).

4. Soil Invertebrates. No data on the toxicity of DNB to terrestrial invertebrates were located.

5. Soil Microorganisms. No data on the effect of DNB on microorganisms in soil were located. In laboratory culture solutions, growth of microorganisms was inhibited at 4–100 mg/L (Kitchens et al. 1978; Wentzel et al. 1979). Using the cell multiplication inhibition test, Bringmann and Kuhn (1980) found a "toxicity threshold" of 14 mg/L for *Pseudomonas putida*. In a synthetic medium, 16 species of microorganisms isolated from soil exposed to DNB waste effluent were capable of degrading DNB present at an initial concentration of 100 mg/L (Dey et al. 1986).

6. Metabolism and Bioaccumulation.

Animals. DNB can enter the mammalian body by ingestion, inhalation, or dermal absorption (ATSDR 1995b). Metabolism of DNB proceeds by enzymic reduction of one or both nitro groups, followed by addition of a hydroxyl group adjacent to one of the resulting amino groups. The metabolites are acetylated or conjugated with glucuronic acid and excreted primarily in the urine. In metabolic studies with rabbits, 65%–93% of orally administered DNB (50–100 mg/kg) was excreted in the urine, primarily as the metabolites 3-nitroaniline, 1,3-phenylenediamine, 2,4-diaminophenol, and 2-amino-4-nitrophenol during the first 2 d (Parke 1961). In rats, reduction followed by acetylation results in the excretion of 3-aminoacetanilide, 4-acetamidophenol, 1,3-diacetamidobenzene, 3-nitroaniline-*N*-glucuronide, and 3-nitroaniline in the urine (Nystrom and Rickert 1987). Sixty-eight percent of the dose was eliminated in 24 hr; 1%–5% of the dose was excreted in the feces. McEuen et al. (1995) additionally identified 2,4-diacetamidophenol and 2-nitro-acetamidophenol but not 3-aminoacetanilide in the urine of orally dosed male rats. Excretion of radiolabeled metabolites in the urine and feces in two of these studies indicated that 63%–93% of an oral dose is absorbed by the gastrointestinal tract of rats and rabbits (Parke 1961; Nystrom and Rickert 1987). Following single oral doses to rats, levels of radioactivity were highest in the liver, white fat, brown fat, kidney, and sciatic nerve (Philbert et al. 1987). Levels were higher in germfree rats than in conventional rats, indicating the importance of intestinal microflora in bioavailability and metabolism.

Because DNB is readily metabolized and excreted in mammalian systems, it is not expected to bioaccumulate. Biomonitoring studies have been conducted at Aberdeen Proving Ground, MD (USACHPPM 1994), and at several other AAPs (see Opresko 1995b for review). DNB was not present in tissues of terrestrial wildlife (deer and small mammals) at or above a detection limit of 0.05 mg/kg.

Plants. McFarlane et al. (1987) studied the uptake, distribution, and metabolism of ^{14}C-labeled DNB in hydroponically grown mature soybean plants. The initial uptake rate, measured by several methods, ranged from 17 to 22 mL/min. Radioactivity remained primarily in the roots, with slow translocation of ^{14}C to the shoots and leaves. The thin-layer chromatography pattern of root extracts indicated that about half the parent chemical was degraded to other compounds within the roots. The authors suggested that ^{14}C in the leaves probably resulted from translocated degradation products although the metabolites were not identified.

E. Terrestrial Criteria and Screening Benchmarks

1. Mammals. A chronic oral RfD for humans of 0.0001 mg/kg/d was calculated by extrapolation and multiplication by uncertainty factors from a subchronic oral study with rats (USEPA 1993d). The endpoint in that study was increased splenic weight.

Three subchronic studies lasting 84–112 d on the oral toxicity of DNB utilizing laboratory animals were available for calculating screening benchmarks for mammalian wildlife (Cody et al. 1981; Linder et al. 1986; Reddy et al. 1995). In all three studies, the relevant endpoint was testicular degeneration, an effect that would diminish population survival. All three studies, using different strains of laboratory rats, showed remarkedly similar effects at similar dose levels (Table 30): a NOAEL ≤1.13 mg/kg/d and a LOAEL ≥1.5 mg/kg/d. The highest NOAEL of 1.13 mg/kg/d from the study by Cody et al. (1981) was chosen for calculation of screening benchmarks for mammalian wildlife. The subchronic NOAEL was multiplied by an uncertainty factor of 0.1 to derive the chronic NOAEL. The time-weighted average body weight of the animals in this test group was not reported; the terminal average body weight was 0.450 kg. To calculate the wildlife benchmarks, the EPA default value of 0.350 kg was used for the body weight of the test species. The NOAEL was extrapolated to seven other species of wildlife as in the TNT section. The log K_{ow} of 1.62 was used to estimate the FCM for the mink. Screening benchmarks for the other species of wildlife are listed in Table 31.

Confidence in the benchmarks is high because of the number of studies and similar results. Additional confidence would be placed in the benchmarks if similar results were attained with a second species.

2. Birds. No subchronic or chronic studies were available for calculation of screening benchmarks for avian species.

3. Plants. Insufficient data were located for calculation of a screening benchmark for terrestrial plants. In one study a solution of 34 µg/L was not phytotoxic (McFarlane et al. 1987), but the growth period was only 4 d.

4. Soil Invertebrates. No data were located for calculation of a screening benchmark for soil invertebrates.

5. Soil Heterotrophic Processes. Insufficient data were located for calculation of a screening benchmark for soil heterotrophic processes. Studies with laboratory cultures utilized acclimated organisms.

V. 3,5-Dinitroaniline

3,5-Dinitroaniline (DNA) is a weak explosive. Nitration of DNA yields the powerful explosive, 2,3,4,5,6-pentanitroaniline. DNA is prepared from TNB or from 3,5-dinitrobenzoic acid (Fedoroff et al. 1960).

DNA is formed during the production of TNT, either through the nitration of toluene (the starting material for production of TNT) or through the chemical or microbial degradation of nitrated toluenes. It is released to the environment from munitions production and processing facilities. It is also formed in the environment by the bacterial reduction of TNB, another by-product of munitions

Table 30. DNB toxicity data for mammalian species.

Species	Route	Exposure period	NOAEL (mg/kg/d)	LOAEL (mg/kg/d)	Effect/endpoint	Reference
Rat	Gavage	12 wk	1.1	2.1	Testes, sperm production	Linder et al. 1986
	Diet	90 d	0.35	1.73	Testicular degeneration	Reddy et al. 1995
	Drinking water	16 wk	1.13	2.64	Testes, sperm production	Cody et al. 1981

Table 31. DNB Screening benchmarks for selected mammalian wildlife species.

Wildlife species	Chronic NOAEL (mg/kg/d)	Diet (mg/kg food)	Water (mg/L)	Piscivorous species (mg/L)[a]
Shorttail shrew	0.25	0.41	1.13	—
White-footed mouse	0.23	1.46	0.75	—
Meadow vole	0.19	1.66	1.40	—
Cottontail rabbit	0.08	0.42	0.86	—
Mink	0.09	0.63	0.84	0.06
Red fox	0.06	0.60	0.71	—
Whitetail deer	0.03	1.02	0.49	—

[a]Water concentration that incorporates dietary intake from both water and food consumption.

manufacturing (Mitchell et al. 1982; Spanggord et al. 1982a). No data on concentrations in the environment were located; however, DNA is usually detected in soils containing TNB (Walsh and Jenkins 1992).

Few data on transport in the environment and no data on toxicity to terrestrial mammals, birds, plants, soil invertebrates, or soil heterotrophic processes were located. Aquatic toxicity data on one species of invertebrate and three species of fish were located. These data were used to derive Tier II screening benchmarks for DNA. Chemical and physical properties are listed in Table 32.

Table 32. Chemical and physical properties of 3,5-dinitroaniline (DNA).

Synonyms	1-Amino-3,5-dinitrobenzene 3,5-Dinitrobenzenamine	HSDB 1995d
CAS number	618-87-1	HSDB 1995d
Molecular weight	183.12	Budavari et al. 1996
Physical state	Yellow needles	Fedoroff et al. 1960
Chemical formula	$C_6H_5N_3O_4$	Budavari et al. 1996
Structure	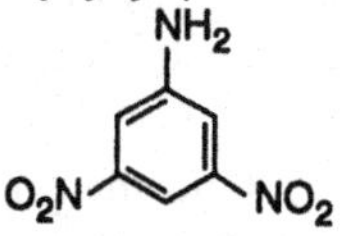	Budavari et al. 1996
Water solubility	Slightly soluble in cold water	Fedoroff et al. 1960
Specific gravity	No data	
Melting point	158–160 °C	Fedoroff et al. 1960
Boiling point	No data	
Vapor pressure (20 °C)	No data	
Partition coefficients		
Log K_{ow}	No data	
Log K_{oc}	2.4–2.7	von Oepen et al. 1991
Henry's law constant (20 °C)	No data	

A. Environmental Fate

1. Sources and Occurrences. DNA is present in condensate wastewater derived from the production and purification of TNT at production facilities such as AAP. It was present in 7.6% of condensate wastewater at concentrations ranging from 0.005 to 0.30 mg/L at an AAP that used the sellite process for manufacture of TNT. In the environment, microbial transformation of TNT results in formation of DNA (Mitchell et al. 1982).

Air. No information on the occurrence of DNA in air was located.

Surface Water and Groundwater. No monitoring studies of DNA in surface water and groundwater at military sites were located.

Soils. DNA was present in soils at the following U.S. Army sites: Weldon Springs, Hawthorne AAP, Hastings East Park, Nebraska Ordnance Works, and Raritan Arsenal, but concentrations were not given. DNA was usually detected in soils containing TNT. Chromatograms indicated that concentrations were low compared to other munitions contaminants (Walsh and Jenkins 1992). Four soil samples taken from the Naval Surface Warfare Center contained concentrations ranging from 0.08 to 0.67 mg/kg (dry weight) (Grant et al. 1995).

2. Transport and Transportation Processes.

Abiotic Processes. No data on transport or transformation in air were located.

No data on transport or transformation in water were located. DNA is only slightly soluble in water (Fedoroff et al. 1960) and two isomers of DNA, 2,4- and 2,6-dinitroaniline, are practically insoluble in water (Budavari et al. 1996). Aromatic amines are generally resistant to aqueous hydrolysis (Lyman et al. 1982).

Anilines strongly bind to humic substances in soil and are not easily extracted (Bollog and Loll 1983). However, measured K_{oc} values for DNA of 507, 303, and 253 for podzol, alfisol, and sediment, respectively (von Oepen et al. 1991), indicate a moderate potential to bind to soil.

Biotransformation. Incubation of TNB in the presence of nutrients results in formation of DNA (Mitchell et al. 1982). The metabolite was formed when TNB was added to nutrient-enriched water samples from the Tennessee River. Incubation of DNA under the same conditions as those for TNB resulted in formation of 3,5-diaminonitrobenzene (the principal product), 3-nitro-5-aminoacetanilide, and a nitro-, amino-substituted *N*-methylindoline.

Mitchell et al. (1982) conducted microbial screening tests using Tennessee River water and sediments collected downstream of the Volunteer AAP. In tests lasting 19 d, concentrations of DNA (14.5 μg/mL) decreased 33%–52% only in water containing sediments; degradation was preceded by a lag period of 12 d.

Cultures enriched in microorganisms capable of biodegrading DNA were not capable of using the compound as a sole carbon source; biodegradation occurred only with the addition of glucose and yeast extract. Tests with ^{14}C-labeled DNA did not result in liberation of $^{14}CO_2$, indicating that complete mineralization did not occur.

Mitchell et al. (1982) also conducted 3-d bioadsorption studies with DNA and viable or heat-killed cells of *Escherichia coli, Bacillus cereus, Serratia marcescens*, and *Azobacter beijerinckii*. Biosorption coefficients ([μg absorbed/g bacteria]/[μg chemical/mL supernatant]) of less than 10 indicated little bioadsorption.

B. Aquatic Toxicology

1. Acute Effects: Invertebrates and Fish. Static acute tests were conducted with one invertebrate and four species of fish (Table 33). The results of tests with daphnids and fish from two different laboratories were very similar. DNA was not highly acutely toxic to aquatic organisms; the most sensitive species was the rainbow trout with a 96-hr LC_{50} of 3.0 mg/L. An 11-d test conducted under flow-through conditions resulted in a similar LC_{50} value for rainbow trout.

An acute toxicity test was also conducted with a saltwater invertebrate, the

Table 33. Acute toxicity of DNA to aquatic invertebrates and fish.[a]

Test species	Test duration	EC_{50}[b] or LC_{50} (mg/L)	Reference
Daphnia magna (water flea)	48-hr	15.4	Liu et al. 1983a
Daphnia magna (water flea)	48-hr	13.8[c]	van der Schalie 1983
Pimephales promelas (fathead minnow)	96-hr	21.8	Liu et al. 1983a
Pimephales promelas (fathead minnow)	96-hr	21.2[c]	van der Schalie 1983
Lepomis macrochirus (bluegill sunfish)	96-hr	7.0[c]	van der Schalie 1983
Oncorhynchus mykiss (rainbow trout)	96-hr 11-d[d]	3.0[c] 2.0	van der Schalie 1983
Ictalurus punctatus (channel catfish)	96-hr	13.9[c]	van der Schalie 1983

[a]All tests were conducted under static conditions except where otherwise noted.
[b]EC_{50} (immobilization) are for *Daphnia magna*.
[c]Measured concentration.
[d]Flow-through test.

shrimp *Crangon septemspinosa*. The 96-hr LC_{50} was 11.6 mg/L (McLeese et al. 1979).

2. Chronic Effects: Invertebrates and Fish. Chronic tests were conducted under flow-through conditions with one species of invertebrate and one species of fish (Table 34). Daphnids were tested at measured concentrations of 1.29 to 7.98 mg/L and rainbow trout were tested at measured concentrations of 0.15 to 1.21 mg/L. The LOECs were 4.56 mg/L for *Daphnia magna* and 0.65 mg/L for *Oncorhynchus mykiss*.

3. Plants. Only one species, the green alga *Selenastrum capricornutum*, was tested. Under static conditions, significantly reduced growth, measured on d 5 of the test, occurred at concentrations of 0.13–15.1 mg/L (van der Schalie 1983). These concentrations were considered algistatic because transfer to clean media resulted in renewed growth. A concentration of 0.03 mg/L was a no-effect concentration.

4. Metabolism and Bioconcentration. No information on metabolism and bioconcentration in aquatic organisms was located.

C. Aquatic Criteria and Screening Benchmarks

1. Aquatic Organisms. Data from acute tests were available for only four of the required eight families: Salmonidae (*O. mykiss*), two additional families in the class Osteichthyes (*Lepomis macrochirus* and *Ictalurus punctatus*), and a planktonic crustacean (*D. magna*). Therefore, Tier II or secondary acute and chronic values were calculated according to USEPA guidance for the Great Lakes System (USEPA 1993a). For two of the families, data were available to calculate acute/chronic toxicity ratios, Daphnidae and Salmonidae.

The SAV was calculated by dividing the lowest GMAV in the database by the SAF. The lowest GMAV was 3.0 mg/L for *O. mykiss*. USEPA (1993a) lists a SAF of 6.5 for use in Tier II calculations when four satisfied data requirements for Tier I calculations are available.

Therefore

$$\text{SAV} = \frac{3.0}{6.5} = 0.46 \text{ mg/L}$$

The SMC is one-half of the SAV or 0.23 mg/L.

For the SACR, two of the three experimentally determined ACRs were available according to guidelines for Tier I. The geometric mean of the two acute values for *D. magna* was used as the acute value. LOECs and NOECs for the chronic studies are listed in Table 34. So that the total number of ACRs is three, an additional default ACR of 18 was added according to USEPA (1993a) guidelines. The SACR is the geometric mean of the three ACRs listed in Table 35, 7.8 mg/L.

Table 34. Chronic toxicity of DNA to aquatic invertebrates and fish.[a]

Species	Stage/age	Parameter measured	Parameter response	Concentration (mg/L)	Reference
Daphnia magna (water flea)	Neonate	Survival and reproduction (21 d)	No effect (NOEC)	2.41	van der Schalie 1983
			Decreased reproduction (LOEC)	4.56	
			Decreased length (LOEC)	4.56	
Oncorhynchus mykiss (rainbow trout)	Egg/fry	Time to hatch, hatching success, fry survival, overall survival, fry deformities, fry length, fry weight (71 d)	No effect (NOEC)	0.37	van der Schalie 1983
			Decreased hatching success and Survival (LOEC)	0.65	

[a]Tests were conducted under flow-through conditions; all concentrations were measured.

Table 35. Aquatic acute/chronic ratios (ACR) for DNA.

Species	Acute values (mg/L)	Chronic values (mg/L)	Acute/chronic ratios (ACR)
Daphnid	14.5	3.32	4.37
Rainbow trout	3.0	0.49	6.12
Default ACR[a]			18

[a]A default ACR of 18 is used when fewer than three ACRs are available (USEPA 1993a).

The SCV is the SAV divided by the SACR:

$$SCV = \frac{SAV}{SACR} = \frac{0.46}{7.8} = 0.059 \text{ mg/L}$$

Aquatic criteria/screening benchmarks are listed in Table 36.

The single test with the green alga, *S. capricornutum*, was used to approximate a final plant value. A lowest-effect value in this test was 0.13 mg/L.

2. Sediment-Associated Organisms. No data on the toxicity of DNA to sediment-associated organisms were located. Although a dissociation constant for DNA was not located, aromatic amines are known to dissociate under environmental conditions (pH, 5–9) (Howard et al. 1989). Because of the potential ionization of the amine group of DNA under ambient conditions, the equilibrium partitioning method of Di Toro et al. (1991) for calculating SQC for nonionic organic chemicals cannot be utilized.

D. Terrestrial Toxicology

1. Toxicity. No data on the toxicity of DNA to mammals, birds, plants, soil invertebrates, or soil microbes were located.

Table 36. Water quality criteria/screening benchmarks for DNA.

Criterion	Value (mg/L)
Acute water quality criterion	Insufficient data
Chronic water quality criterion	Insufficient data
Secondary acute value	0.46
Secondary maximum concentration	0.23
Secondary chronic value	0.06
Secondary continuous concentration	0.06
Lowest chronic value, fish	0.65
Lowest chronic value, daphnids	4.6
Final plant value	0.13
Sediment quality benchmark	Insufficient data

2. Metabolism and Bioconcentration.

Animals. No data on the metabolism and bioaccumulation of DNA were located. However, the disposition of the related isomer 2,4-dinitroaniline, administered to rats, was studied. The compound was rapidly absorbed, distributed to all major tissues within 45 min, rapidly cleared, and excreted in the urine (63% in 24 hr) and feces (23% in 3 d). 2,4-Dinitrophenylhydroxylamine was the main metabolite, with excretion in the urine as the sulfate conjugate and in bile as the glucuronide (Matthews et al. 1986). By analogy, hydroxylation of the amino group of DNA would result in excretion of 3,5-dinitrophenylhydroxylamine. Rapid metabolism indicates no bioaccumulation.

Plants. No data on the metabolism or bioaccumulation of DNA by plants were located.

E. Terrestrial Criteria and Screening Benchmarks

No data useful for calculating criteria or screening benchmarks for terrestrial wildlife, birds, plants, or soil invertebrates or heterotrophic processes were located.

VI. 2-Amino-4,6-Dinitrotoluene

2-Amino-4,6-dinitrotoluene (2-ADNT) is formed during the production of TNT and is released to the environment from munitions production and processing facilities. It is also formed from TNT present in the environment by bacterial reduction of the 2-nitro group. Chemical and physical properties are listed in Table 37.

Few data on transport and transformation processes in the environment were located. 2-ADNT is moderately persistent in the environment; important fate mechanisms include photolysis and biodegradation. Data from acute toxicity tests with two species of aquatic organisms were used to derive Tier II acute and chronic screening benchmarks. No data on toxicity to terrestrial mammalian species, soil invertebrates, or soil heterotrophic processes were located. Studies were located on terrestrial plant uptake and phytotoxicity.

A. Environmental Fate

1. Sources and Occurrences. 2-ADNT is present in condensate wastewater released to the environment from the production and purification of TNT at AAPs. According to Spanggord et al. (1982a), it most likely arises from the bacterial reduction of TNT. It is also formed in the environment from the bacterial reduction of the 2-nitro group of TNT (McCormick et al. 1976).

Air. No data on concentrations in air were located.

Table 37. Chemical and physical properties of 2-amino-4,6-dinitrotoluene (2-ADNT).

Synonyms	3,5-Dinitro-*o*-toluidine	RTECS 1995
	2-Methyl-3,5-dinitrobenzenamine	
CAS number	35572-78-2	RTECS 1995
Molecular weight	197.17	RTECS 1995
Physical state	Crystals	Kaye 1980a
Chemical formula	$C_7H_7N_3O_4$	RTECS 1995
Structure	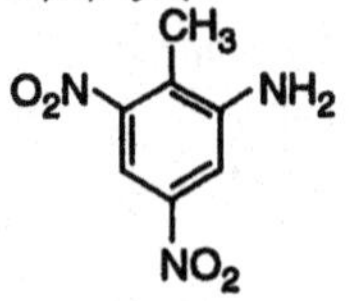	McCormick et al. 1976
Water solubility (20 °C)	2800 mg/L (estimated)	Layton et al. 1987
Specific gravity (20 °C)	No data	
Melting point	173–176 °C	Kaye 1980a
Boiling point	No data	
Vapor pressure (20 °C)	4×10^{-5} torr (estimated)	Layton et al. 1987
Partition coefficients		
Log K_{ow}	1.06 (estimated)	Liu et al. 1983a
	1.94 (estimated)	Jenkins 1989
	0.5 (estimated)	Layton et al. 1987
K_d	4	Pennington and Patrick 1990
Henry's law constant (20 °C)	3×10^{-3} L-torr/mol (estimated)	Layton et al. 1987

Surface Water and Groundwater. 2-ADNT occurred in 21.5% of condensate wastewater samples at a TNT production facility at an average concentration of 0.02 mg/L (range, 0.001–0.10 mg/L) (Pearson et al. 1979; Spanggord et al. 1982a). 2-ADNT was detected in groundwater at the Hawthorne Naval Ammunition Depot, but concentrations were not given (Pereira et al. 1979). Water samples downstream of the Holston AAP were not analyzed for 2-ADNT, but concentrations in Holston River sediment were 0.0675 mg/L near a TNT discharge line and 0.0004 mg/L at a site approximately 8 km downstream (Spanggord et al. 1981). The metabolic precursor, 2-hydroxylamino-4,6-dinitrotoluene (McCormick et al. 1976), was detected in water and sediment samples at the Iowa AAP (Sanocki et al. 1976).

Soils. At the Joliet AAP, 2-ADNT was detected in soil samples collected from a LAP area and an open burning ground. Concentrations ranged from less than the limit of detection (0.1 mg/kg) in approximately half the samples to 19 mg/kg (Simini et al. 1995). In a survey of TNT and transformation products in a limited number of soil samples collected at AAP, depots, and arsenals, the following concentrations (air-dried weight) were found: Nebraska Ordnance Works, 0.05–8.0 mg/kg; Newport, 0.3 mg/kg; Weldon Springs, 0.05–2.3

mg/kg; Raritan Arsenal, 2.0–37 mg/kg; Hastings East Park, 0.05 mg/kg; Sangamon Ordnance Plant, 4.9 mg/kg; and Eagle River Flats, 0.05–0.73 mg/kg (Walsh and Jenkins 1992). At nine other sites, samples were not analyzed or concentrations were below the limit of detection.

2. Transport and Transformation Processes.

Abiotic Processes. No data on transformation in the atmosphere were located.

Burlinson and Glover (1977) reported that irradiation of 50 mg/L of 2-ADNT in an unbuffered aqueous solution at >290 nm produced a 70% loss in 7 hr. Six products were isolated by thin-layer chromatography, but only 4,6-dinitroindazole was identified.

Soil partition coefficients (K_d) were measured in a soil–water system containing TNT and its metabolites. Under both oxidized and reduced conditions, K_d for 2-ADNT were similar, 4.9 and 3.7, respectively (Pennington and Patrick 1990). A K_d of approximately 4 indicates little sorption to soil.

Biotransformation. Because 2-ADNT is a biodegradation product of TNT, its biotransformation has been studied in association with the biodegradation of TNT. 2-ADNT is formed from TNT by the reduction of the 2-nitro group to an amino group. Further reduction leads to formation of 2,4-diamino-6-nitrotoluene, 2,6-diamino-4-nitrotoluene, and triaminotoluene (Hoffsommer et al. 1978; McCormick et al. 1976) (see Fig. 1). The compound is formed in cultures of TNT incubated with *Pseudomonas* sp. (Won et al. 1974) and with microbial populations isolated from sludge (Hoffsommer et al. 1978) and sediment (Spanggord et al. 1980b). In batch studies, incubation of 13 mg/L of 2-ADNT with sludge microbes for 3 d resulted in complete metabolism to 2,4-diamino-6-nitrotoluene and 2,6-diamino-4-nitrotoluene (Hoffsommer et al. 1978).

2-ADNT is moderately persistent in soil. Cataldo et al. (1989) amended various soils with unlabeled or ^{14}C-labeled-TNT and isolated the degradation products. The soils ranged from sandy to silt loams with organic carbon contents of 0.5% to 7.2%. 2-ADNT and 4-amino-2,6-dinitrotoluene (as determined by HPLC chromatograms and mass spectrometry) appeared almost immediately in the soil and increased dramatically by d 10. After 60 d of incubation the two isomers accounted for >80% of the TNT-derived activity in two of the soils. Similar but slower transformation occurred in radiation-sterilized soils, indicating that abiotic processes may account for some or all of the transformation.

Two types of soil, a Tunica silt and a Sharkey clay, were amended with 2-ADNT and extracted for metabolites after 20 and 65 d (Pennington 1988a,b). In addition to the parent compound, the following metabolites were present in both soils at both times: TNT, 4-amino-2,6-dinitrotoluene, 2,6-diamino-4-nitrotoluene, 2,4-diamino-6-nitrotoluene, 4-amino-2-nitrotoluene, 2,4-dinitrotoluene, 2,6-dinitrotoluene, and 1,3,5-trinitrobenzene. After 65 d, 2-ADNT persisted in the soils in greater concentrations than the other metabolites. The sum of the recoveries of all the compounds was low compared to the original spike (80 mg/kg of soil, oven dry weight basis). The author suggested that spontaneous reduction could have occurred in the soils or during the extraction procedure.

Using the white-rot fungus *Phanerochaete chrysosporium*, immobilized in a fixed-film silicone membrane bioreactor, TNT was rapidly degraded (half-life, 4.4 hr) compared to 2-ADNT (half-life, 50 hr), suggesting that degradation of aminodinitrotoluene isomers is the rate limiting step in biodegradation of TNT (Bumpus and Tatarko 1994). 2-ADNT was degraded to an unidentified water-soluble metabolite. Under carefully controlled anaerobic conditions, TNT was degraded by soil microbes to the 2- and 4-aminodinitrotoluenes within 10 d and the aminodinitrotoluene isomers were removed within 25 d (Funk et al. 1993). The bacterium *Pseudomonas aeruginosa* (strain MA01), cultured with an exogenous carbon source, cometabolized 2-ADNT into polar metabolites (Alvarez et al. 1995). Approximately 50% of the 2-ADNT was metabolized to polar metabolites in 170 hr; other metabolites were identified as the diaminonitrotoluenes, acetyl-diaminonitrotoluenes, and tetranitroazoxytoluenes.

B. Aquatic Toxicology

1. Acute Effects: Invertebrates and Fish. Data were available on one species of freshwater invertebrate and one species of fish. In static tests, the 48-hr EC_{50} for immobilization of *Daphnia magna* was 4.6 mg/L; the 96-hr LC_{50} for *Pimephales promelas* was 15.1 mg/L (Bailey and Spanggord 1983; Liu et al. 1983a).

Won et al. (1976) evaluated the acute toxicity of 2-ADNT to two saltwater species, a tidepool copepod, *Tigriopus californicus*, and oyster larvae, *Crassostrea gigas*. Groups of 100 copepods and 200 oyster larvae (average age, 20 d) were incubated at 20 °C for 72 and 96 hr, respectively, and evaluated for mortality. No mortality was observed for either species at the highest concentration tested, 50 mg/L.

2. Chronic Effects: Invertebrates and Fish. No data on chronic exposures of aquatic organisms were located.

3. Plants. Won et al. (1976) evaluated the toxicity of 2-ADNT to the freshwater green alga *Selenastrum capricornutum*. Incubation was for 7 d and toxicity was evaluated in terms of dry weight and morphological alterations. No toxicity was observed at the highest concentration tested, 50 mg/L.

4. Metabolism and Bioconcentration. Using reverse-phase HPLC retention time, Jenkins (1989) calculated a log K_{ow} of 1.94. From an estimated log K_{ow} of 1.06, Liu et al. (1983a) calculated a bioconcentration factor of 3.76, which indicates little potential for bioaccumulation.

C. Aquatic Criteria and Screening Benchmarks

1. Aquatic Organisms. Data were insufficient for calculation of acute or chronic WQC according to USEPA guidelines. Although only two acute toxicity studies were available, these data are sufficient to calculate acute and chronic Tier II criteria according to USEPA (1993a) guidelines for the Great Lakes System. Data were available on a planktonic crustacean (*D. magna*) and a family in the class Osteichthyes (Cyprinidae, *P. promelas*). No chronic data were available for calculation of acute/chronic toxicity ratios.

The SAV was calculated by dividing the lowest GMAV in the database by the SAF. The lowest GMAV was 4.6 mg/L for *D. magna*. USEPA (1993a) lists a SAF of 13 for use in Tier II calculations when two satisfied data requirements for Tier I calculations are available.

Therefore

$$\text{SAV} = \frac{4.6}{13} = 0.35 \text{ mg/L}$$

The SMC is one-half of the SAV or 0.18 mg/L.

No experimentally determined acute/chronic ratios were available for determination of the SACR; therefore, the default SACR of 18 was used. The SCV is the SAV divided by the SACR:

$$\text{SCV} = \frac{\text{SAV}}{\text{SACR}} = \frac{0.35}{18} = 0.019 \text{ mg/L}$$

The single test with the green alga *S. capricornutum* was used to approximate a FPV. The FPV would be the same as the lowest chronic value for aquatic plants, >50 mg/L. Aquatic screening benchmarks are listed in Table 38.

2. Sediment-Associated Organisms. No data on sediment-associated organisms were located. Calculation of SQB based on EqP (Di Toro et al. 1991) applies to nonionic organic chemicals, and a SQB cannot be calculated for 2-ADNT because of the potential ionization of the amine group under environmental conditions.

Table 38. Water quality criteria/screening benchmarks for 2-ADNT.

Criterion/Benchmark	Value
Acute water quality criterion	Insufficient data
Chronic water quality criterion	Insufficient data
Secondary acute value	0.35 mg/L
Secondary maximum concentration	0.18 mg/L
Secondary chronic value	0.02 mg/L
Secondary continuous concentration	0.02 mg/L
Lowest chronic value, fish	Insufficient data
Lowest chronic value, daphnids	Insufficient data
Final plant value	>50 mg/L
Sediment quality benchmark	Insufficient data

D. Terrestrial Toxicology

1. Mammals. No studies on the subchronic or chronic administration of 2-ADNT to wildlife or laboratory-reared mammals were located.

2. Birds. No data on the toxicity of 2-ADNT to birds were located.

3. Plants. Yellow nutsedge (*Cyperus esculentus*) was grown for 42 d in clay or silt soil amended with 80 μg of 2-ADNT/g of soil (80 mg/kg, oven dry weight basis) (Pennington 1988a,b). There were no significant differences in yield, as measured by plant weight, between controls and treatments within soil types.

4. Soil Invertebrates. No data on the toxicity of 2-ADNT to soil invertebrates were located.

5. Soil Heterotrophic Processes. No studies designed to evaluate the toxicity of 2-ADNT to soil microorganisms were located. In biotransformation studies, no toxic effects on sludge microorganisms were reported at a treatment concentration of 13 mg/L (Hoffsommer et al. 1978) or on soil microorganisms at a treatment concentration of 80 mg/kg (soil dry weight) (Pennington 1988a,b).

6. Metabolism and Bioaccumulation.

Animals. No data on the metabolism of 2-ADNT in mammalian species were located. The compound has been found in the urine of workers engaged in the production of TNT and in the urine of laboratory animals administered TNT orally (El-hawari et al. 1981; Hodgson et al. 1977) or dermally (Yinon and Hwang 1987). 2-ADNT is excreted primarily as a glucuronide conjugate. Further reduction to 2,4-diamino-6-nitrotoluene and 2,6-diamino-4-nitrotoluene may take place before excretion as these metabolites are also found in the urine of mammalian species administered TNT (ATSDR 1995a).

No information on bioaccumulation was located. Based on the administration of TNT to mammalian species, the compound appears to be metabolized and excreted in the urine. Biomonitoring studies have been conducted at Aberdeen Proving Ground, MD (USACHPPM 1994), and at several other AAPs (see Opresko 1995b for review). 2-ADNT was not present in tissues of terrestrial wildlife (deer and small mammals) at or above a detection limit of 0.1 mg/kg.

Plants. Yellow nutsedge grown for 42 d in hydroponic cultures containing TNT and analyzed for TNT and its metabolites contained primarily the metabolites 4-amino-2,6-dinitrotoluene and 2-ADNT (Palazzo and Leggett 1986). Highest concentrations were present in the roots; at 20 mg/L of TNT, the roots contained almost 2200 mg/kg of 4-amino-2,6-dinitrotoluene and 600 mg/kg of 2-ADNT. The metabolites were found throughout the plants. Because no metabolites were present in the culture solution (the containers were painted black to inhibit photolysis), the authors concluded that metabolism took place within the plant. No other metabolites were reported.

Neither the parent compound nor metabolites could be detected in yellow nutsedge grown in soil amended with 80 mg/kg of 2-ADNT (Pennington 1988a,b). Plants were grown in pots containing clay or silt soil for 45 d before harvesting. The authors suggested that bioavailability was limited by volatilization of the compound or its metabolites and adsorption of the compounds or its metabolites to soil, but data on volatilization and adsorption were not available.

A limited number of uptake and metabolism studies indicate that 2-ADNT does not bioaccumulate in plants, either due to metabolism or lack of uptake.

E. Terrestrial Criteria and Screening Benchmarks

1. Mammals. No studies on the subchronic or chronic toxicity of 2-ADNT to mammals or birds were located; thus, screening benchmarks could not be calculated.

2. Birds. No subchronic or chronic studies were available for calculation of screening benchmarks for avian species.

3. Plants. In the absence of criteria for terrestrial plants, LOECs (or NOECs) from the literature can be used to screen for chemicals of potential concern (Will and Suter 1995a). As indicated in a metabolism study, soil amended with 80 mg/kg of 2-ADNT was not toxic to yellow nutsedge over a 42-d period (Pennington 1988a,b). Based on this single study, confidence in this screening benchmark (Table 39) is low.

4. Soil Invertebrates. No studies on the toxicity of 2-ADNT were located; thus, a screening benchmark could not be calculated.

5. Soil Heterotrophic Processes. In the absence of criteria for soil heterotrophic processes, LOECs (or NOECs) from the literature can be used to screen

Table 39. 2-ADNT screening benchmarks for terrestrial plants and invertebrates.

Screening benchmark	Value
Plants, solution	Insufficient data
Plants, soil	80 mg/kg[a]
Soil invertebrates (earthworms)	Insufficient data
Soil invertebrates (other)	Insufficient data
Soil microbial processes	80 mg/kg[a]

[a]No-effect concentration.

for chemicals of potential concern (Will and Suter 1995b). A soil concentration of 80 mg/kg was reported as nontoxic to microbial processes (Pennington 1988a). However, the organisms in the study were probably acclimated to the test compound. Therefore, this potential screening benchmark should not be considered conservative and, because it is based on a single study, confidence in the benchmark is low.

VII. Hexahydro-1,3,5-Trinitro-1,3,5-Triazine

Hexahydro-1,3,5-trinitro-1,3,5-triazine (RDX) is a crystalline high explosive used extensively by the military in shells, bombs, and demolition charges. It is commonly referred to as cyclonite or RDX (British code name for Research Department Explosive or Royal Demolition Explosive) (Fedoroff and Sheffield 1966). Other synonyms are hexolite, cyclonite, hexogen, PBX (AF) 108, T4, cyclotrimethylenetrinitramine, trimethylene-trinitramine, and trinitrocyclotrimethylene (HSDB 1995e). Manufacture in the U.S. is by the Bachmann process in which hexamine is reacted with an ammonium nitrate/nitric acid mixture in the presence of acetic acid and acetic anhydride (Yinon 1990). Military grades of RDX contain about 10% octahydro-1,3,5,7-tetranitro-1,3,5,7-tetrazocine (HMX). Chemical and physical properties are summarized in Table 40.

Information on concentrations in environmental media; environmental fate and transport; and metabolism, bioaccumulation, and effects on aquatic and terrestrial biota were located. Ecotoxicological criteria and screening benchmarks for both aquatic and terrestrial biota, based on population growth and survival effects, were derived.

A. Environmental Fate

1. Sources and Occurrences. RDX is a military-unique compound that is released to the environment at AAPs where it is manufactured. It is also released to the environment during conversion to munitions through a LAP process, at military depot facilities, and through the demilitarization of obsolete munitions. RDX is a persistent chemical in the environment and can be found in soil, surface water, sediments, and groundwater at AAPs and other military sites. As of 1992, ATSDR reported that RDX had been identified at 16 hazardous waste sites proposed for inclusion on the National Priorities List (ATSDR 1995c).

In 1977, four facilities produced RDX, with the primary site of manufacture at Holston AAP, Kingsport, TN (USEPA 1989). As of 1989, Holston AAP was the only U.S. facility engaged in the manufacture of RDX (Burrows et al. 1989). Manufacture and LAP operations were reported at the Cornhusker AAP during the Korean War (Spalding and Fulton 1988) and at the Iowa, Milan, Louisiana, and Lone Star AAPs during the 1970s (Ryon et al. 1984). In the 1970s, LAP operations were carried out in which RDX/HMX formulations were loaded into bombs and projectiles at 11 Army and Navy ammunitions plants (Sullivan et al. 1979). RDX-containing wastewaters are generated at both manufacturing and

Table 40. Chemical and physical properties of hexahydro-1,3,5-trinitro-1,3,5-triazine (RDX).

Synonyms	sym-Trimethylenetrinitramine Cyclotrimethylenetrinitramine 1,3,5-Trinitrohexahydro-*s*-triazine T_4, hexogen Cyclonite 1,3,5-Trinitro-1,3,5-triazacyclohexane	Budavari et al. 1996 HSDB 1995e
CAS number	121-82-4	HSDB 1995e
Molecular weight	222.26	Budavari et al. 1996
Physical state	Orthorhombic crystals from acetone	Budavari et al. 1996
Chemical formula	$C_3H_6N_6O_6$	Budavari et al. 1996
Structure	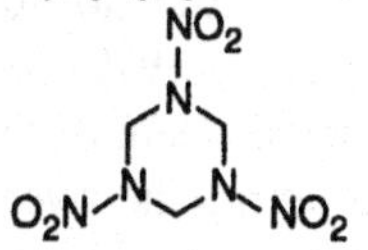	Budavari et al. 1996
Water solubility (20 °C)	Practically insoluble	Budavari et al. 1996
	38.4 mg/L	Spanggord et al. 1983
Specific gravity	1.82	Budavari et al. 1996
Melting point	205°–206 °C	Budavari et al. 1996
Boiling point	Decomposes during melting	Layton et al. 1987
Vapor pressure	1.0×10^{-9} mm Hg	Layton et al. 1987
	4.0×10^{-9} mm Hg	Burrows et al. 1989
Partition coefficients		
Log K_{ow}	0.87	Banerjee et al. 1980
	0.81, 0.86	Burrows et al. 1989
Log K_{oc}	0.88–2.4	Sikka et al. 1980
	2	Burrows et al. (1989)
	1.6, 2.1	Spanggord et al. 1980b
	2.1	Tucker et al. 1985
Henry's law constant	1.2×10^{-5} atm-m^3/mole	McKone and Layton 1986

LAP plants. The major routes of environmental release are discharge in waste streams generated during manufacture and processing, leaching from wastewater storage lagoons or waste burial areas into groundwater, and demilitarization operations. Sampling studies were conducted at various sites over a number of years by different investigators. Following are some of the available data taken from published and unpublished reports.

Air. No data on the presence of RDX in outdoor air were located. A very low vapor pressure, 1.0×10^{-9} mm Hg (Layton et al. 1987) to 4.0×10^{-9} mm Hg

(Burrows et al. 1989), indicates a low potential to enter the atmosphere or to be dispersed by volatilization.

Surface Water, Groundwater, and Sediment. RDX can be released to surface water in wastewater discharges from production and LAP facilities and through overflow of waste lagoons and storage areas. Although of low to moderate water solubility, RDX also has a low adsorption coefficient for soils and sediments and thus will migrate into groundwater.

In 1980, the amount of RDX discharged from the Holston AAP was estimated to be 69.5 kg/d (Spanggord et al. 1980b). Concentrations of nitramines (RDX + HMX) in effluents averaged 2–6 mg/L (Burrows et al. 1989). RDX concentrations in untreated wastewaters ranged from <0.05 (the limit of detection) to 4.75 mg/L, whereas treated wastewaters contained <0.05–0.7 mg/L (Stillwell et al. 1977). Concentrations were measured in the Holston River downstream of the plant by several investigators. Concentrations ranged from 0.7 mg/L immediately below the plant outfall to <0.005–0.07 mg/L approximately 1 mile downstream (Sullivan et al. 1979). Stidham (1979) reported a concentration of 0.079 mg/L 1 mile downstream of the last plant effluent. Concentrations measured at transect lines across the river at 0.1, 4.5, 8, and 33 km were 0.4–224, 3–24, 4–17, and 4–8 μg/L, respectively (Spanggord et al. 1981). The data indicate that, at the initial point of discharge, most of the effluent remained close to the river bank (224 μg/L) with mixing occurring downstream resulting in dilution and smaller ranges of concentrations. In the same study, the sediment concentration at 0 km was 358 ng/g (μg/kg); concentrations in sediment were below the limit of detection at the downstream sites.

Concentrations entering a stream on the Milan AAP site ranged from 0.1 to 109 mg/L with a mean of 11.9 mg/L (Spanggord et al. 1978). Stream water concentrations at the same site ranged from <0.4 to 110 μg/L with sediment concentrations of 290–43,000 mg/kg (Envirodyne Engineers, Inc. 1980). Concentrations in a pink water lagoon at the Milan AAP ranged from <4 to <1600 μg/L with sediment concentrations of 2600–38,000 mg/kg (Envirodyne Engineers, Inc. 1980). Groundwater samples taken below soil containing 0.05–83 mg/kg showed concentrations in 4 of 39 samples of <20 to 780 μg/L (Envirodyne Engineers, Inc. 1980). The maximum concentration found in groundwater at this site was 30.0 mg/L (Tucker et al. 1985). Effluent concentrations at the Iowa AAP ranged from 0.1 to 24 mg/L with a mean of 1.5 mg/L (Spanggord et al. 1978), and concentrations in Brush Creek ranged from 0.1 to 15 mg/L (Small and Rosenblatt 1974). The maximum concentration found in groundwater at the Iowa AAP was 36.0 mg/L (Tucker et al. 1985).

RDX concentrations in water from an inactive lagoon at the Louisiana AAP ranged from 5.6 to 28.9 mg/L; sediment below several lagoons contained 400–120,000 mg/kg, with the highest concentration within the top 0.05 m (Spanggord et al. 1983). Concentrations in groundwater underlying several areas were <1.0–14,120 μg/L (Area P leaching pits), <1.0–707 μg/L (burning ground #5), 8.3 μg/L (landfill #3), and <1.0–537 μg/L (landfill area #8) (Gregory and Elliott

1987). Groundwater below several waste disposal areas was also sampled for RDX by Todd et al. (1989). RDX was detected in 6 of 11 wells below an area of pink water leaching lagoons; concentrations ranged from 13 to 27,000 µg/L. The maximum concentration found in groundwater at the Louisiana AAP was 17.8 mg/L (Tucker et al. 1985).

Spalding and Fulton (1988) summarized the occurrences of several munition residues in the groundwater east of the Cornhusker AAP near Grand Island, NB. In a previous study cited in this paper, onsite groundwater downgradient of the site contained about 300 µg/L RDX; the offsite plume had concentrations >35 µg/L. In 1988, the plume of RDX-contaminated groundwater was 6.5 km long and 1.6 km wide with offsite concentrations up to 100 µg/L. The estimated transport velocity was 0.5 m/d. Concentrations in sediments taken from leaching pits at the site ranged from 2 to 40 mg/kg (Rosenblatt 1986).

Concentrations at the surface of a dry lagoon at the Savanna Army Depot ranged from 3000 to 4000 mg/kg (Rosenblatt 1981). At the Lone Star AAP, sludge below pink water settling ponds contained up to 50,000 mg/kg (Phung and Bulot 1981). Concentrations of <20 to >700 µg/L were found in groundwater in the vicinity of contaminated sites by Burrows et al. (1989). Concentrations in sediment below a pondlike structure used to trap washwater from a U.S. Navy facility (Kitsap County, WA) where projectiles were cleaned ranged from less than the limit of detection to 5.5 mg/kg; concentrations in the water table below the sediments ranged from less than the limit of detection to 5 mg/L (Goerlitz 1992). The facility was active from 1966 to 1970 and the survey was taken in 1974. At an unidentified munitions disposal site, RDX was detected in 7 of 7 groundwater wells at concentrations of 1–47 µg/L (Richards and Junk 1986). It was present in groundwater from a water supply well at an unspecified AAP at a concentration of 70 µg/L (Jenkins et al. 1986).

RDX was not detected in ocean waters near ocean dumping sites for waste munitions (Hoffsommer and Rosen 1972) or in ocean floor sediments near dumping sites (Hoffsommer et al. 1972).

Soils. Contamination of soil results from spills (mainly at military facilities), internment of RDX contaminated wastes in landfills, and open burning/detonation operations.

Over a 10-yr period, Walsh and Jenkins (1992) collected and dried soil samples from 17 U.S. Army sites and analyzed the samples for a number of chemicals relating to munitions use and production. Levels of RDX in excess of the detection limit (0.5 ppm) were found at 11 of these sites; levels ranged from 0.5–1247 mg/kg (median, 19.5) at Nebraska Ordinance Works, 0.5–12,203 mg/kg (median, 38.6) at Newport, 0.5 mg/kg (1/29 samples) at Weldon Springs Training Area, 97.4–13,900 mg/kg (median, 7000) at Iowa AAP, 0.5–4.38 mg/kg (median, 2.4) at Raritan Arsenal, 2.6–8112 mg/kg (median, 127) at Hawthorne AAP, 0.5 mg/kg (1/24 samples) at Hastings East Park, 139–616 mg/kg (median, 378) at Milan AAP, 185–972 mg/kg (median, 578) at Louisiana AAP, and 0.5–0.05 mg/kg at Eagle River Flats explosive ordnance disposal and impact area, to 0.5–3.83 mg/kg (median, 2.2) at Camp Shelby explosive ordnance disposal site and impact area.

As part of a study to evaluate the toxicity of Joliet AAP soils, Simini et al. (1995) determined RDX concentrations in soil taken from LAP and munition burning areas. RDX concentrations ranged from less than the limit of detection (0.1 mg/kg) to 3574 mg/kg, with levels exceeding the detection limit in 13 of 40 samples. Concentrations at open burning sites at the Holston AAP reached 70–80 mg/kg of soil (Bender et al. 1977). Funk et al. (1993) reported that soil samples from an Army munitions depot near Umatilla, OR, contained 3000 mg RDX/kg soil. A soil sample taken from an ordnance-burning area at the Milan AAP contained 39 mg RDX/kg soil (Jenkins and Grant 1987). Four soil samples taken from the Naval Surface Warfare Center contained concentrations ranging from below the limit of detection to 3.3 mg/kg (dry weight) (Grant et al. 1995).

Newell (1984) surveyed open burning and open detonation sites at several AAPs and depots. Concentrations in residues and soil (surface to 18 in. in depth) at the Holston, Iowa, Kansas, Louisiana, Ravenna, Fort Wingate, and Milan AAPs and the Picatinny Arsenal ranged from 1200 to 74,000 mg/kg, with highest concentrations in the residue and soil surface.

2. *Transport and Transformation Processes.*

Abiotic Processes. Because of its low vapor pressure, any RDX in the atmosphere would exist in the particulate form (USEPA 1989). Although RDX was not reported in air samples, any RDX entering the atmosphere would be rapidly degraded. Photolysis of RDX is expected to be an important fate process in the atmosphere since RDX absorbs ultraviolet wavelengths between 240 and 350 nm (Etnier 1986) and it undergoes rapid photolysis in water (Sikka et al. 1980; Spanggord et al. 1980b). RDX in the atmosphere may also be degraded by reaction with photochemically generated hydroxyl radicals (Atkinson 1987).

With a reported solubility of 38.4 mg/L at 20 °C (Spanggord et al. 1983), RDX has low to moderate solubility in water. An estimated Henry's law constant of 1.2×10^{-5} atm-m^3/mole (McKone and Layton 1986) indicates slow to moderate volatilization from water. At neutral or acidic pH values, RDX does not hydrolyze to an appreciable extent in freshwater or seawater. Reported hydrolysis half-lives of 170 d to several years indicate that hydrolysis is not a significant fate mechanism in natural waters (Sikka et al. 1980; Spanggord et al. 1980a). Approximately 12% of a 56 mg/L solution in seawater (pH approximately 8) degraded in 112 d (Hoffsommer and Rosen 1973).

Photolysis is the primary abiotic degradation mechanism in translucent waters. Spanggord et al. (1980b) measured half-lives for RDX (1.3 mg/L) of 13, 14, and 9 d in distilled water, Holston River water, and pond water, respectively, under natural winter conditions in Menlo Park, CA. During a sunny period in March, the half-life of RDX in distilled water was 1.8 d. The half-life was 12 d for RDX in a 4-cm-deep dish of water from a lagoon at the Louisiana AAP (Spanggord et al. 1983). The presence of photoproducts and humic substances did not accelerate the photolysis rate. In another study, RDX in wastewater (23.9 mg/L) exposed to ultraviolet radiation decomposed with a half-life of 3.7 min (Burrows et al. 1984), while photolysis of an aqueous solution of RDX in natural sunlight decomposed with an experimental half-life of 9–13 hr. Similar experiments by Sikka et al. (1980) confirmed the half-life of less than 1 d in surface water exposed to sunlight. Consequently, RDX is not expected to persist for a long period of time in surface waters. Photolysis results in the formation of formaldehyde, nitrite and nitrate ions, and nitroso compounds (Kubose and Hoffsommer 1977; Spanggord et al. 1980b; Burrows et al. 1989).

In a simulation study of the fate of RDX in the Holston River, the major fate process affecting persistence was expected to be dilution (Spanggord et al. 1980b). The photolysis half-life was estimated at 3 d on a sunny day and up to 14 d in winter. Concentrations in the Tennessee River at Knoxville where the Holston River joins the Tennessee River were estimated at 1–3 µg/L.

Sikka et al. (1980) measured the adsorption of RDX on three sediment types and reported partition coefficients (K_p) of 0.80 for sandy loam, 3.06 for clay loam, and 4.15 for organic clay. Calculated K_{oc} values for the respective sediment types were 7.7, 74.6, and 266.7. Even with such low adsorption, steady-state levels of RDX between 30 and 40 ppm were reported for organic and clay sediments in laboratory evaluations.

Tucker et al. (1985) conducted soil sorption experiments with 12 soils of low organic carbon content. K_p values were estimated from regression equations, with independent variables consisting of the organic content of the soil, cation-exchange capacity, pH, and silt content. They found that an organic content of at least 0.25% by weight was necessary for adsorption to be an important process affecting migration. The authors derived a K_{oc} value of 136 from the measurements with the different soils. Using sterilized sediments taken from the Holston River, Spanggord et al. (1980b) calculated adsorption coefficients (K_p) of 4.2 and 1.4 and associated K_{oc} values of 127 and 42. Burrows et al. (1989) reported a K_{oc} of 100 for RDX in soil (type unspecified).

The transport of pink water compounds including RDX (30 mg/L) through columns of garden soil (6.5% organic matter content) amended with microbes from activated sludge and anaerobic sludge digest was studied by Greene et al. (1984). Pink water solutions were continuously pumped through the columns and effluent samples were collected weekly for 110 d. Flow rates were varied and some columns were amended with glucose. In columns not amended with glucose, RDX was not bound and rapidly appeared in leachates. Some degradation, as indicated by decreased recovery, took place in columns with slow flow rates and those amended with glucose.

Soil migration was also studied utilizing both 152-cm-deep × 90-cm-wide cylinders and 5-cm × 61-cm-deep lysimeters, the latter containing ^{14}C-labeled RDX, in a controlled greenhouse experiment (Hale et al. 1979; Kayser and Burlinson 1988). The soils (four types ranging from silty clay to sandy loam with organic carbon content ranging from 0.39% to 2.2%) were irrigated regularly. After 6 mon, concentrations in the large cylinders ranged from the treatment level of 20,000 mg/kg in the upper 10 cm to 150 mg/kg at the 60- to 90-cm depth. RDX was also found at the bottom of the columns. Levels of RDX in water leachates from both experiments were <5 mg/L except in the lysimeters containing coarse, loamy soil where they ranged up to 40 mg/L. Measured K_d values for the four soils were 7.8, 6.4, 0.2, and 1.8 (Hale et al. 1979).

Checkai et al. (1993) collected intact soil-core columns from an uncontaminated area at the Milan AAP to study transport and transformation of munitions chemicals in site-specific soils. The soil was a Lexington silt loam with a 6-in. A horizon containing 16 g/kg organic matter and a B horizon extending from 6 to 27 in. and containing 5 g/kg organic matter. A mixture of munitions simulating open burning/open detonation ash was added to the soil surface. Concentrations were 1000 mg/kg RDX, 1000 mg/kg HMX, 1000 mg/kg 2,4-dinitrotoluene, and 400 mg/kg 2,6-dinitrotoluene. The columns were leached with simulated rainfall for 32.5 wk; controlled tension was applied. RDX was measurable in leachates throughout the study and averaged 12 mg/L. RDX was transported so rapidly in the soil column that it was found at all depths by the first sampling at 6.5 wk.

The foregoing studies indicate that adsorption to soil is not a significant fate mechanism for RDX; likewise, sediment sorption will not lead to significant loss from the aquatic environment. The geometric mean of the K_d values from the studies of Hale et al. (1979), Spanggord et al. (1980b), Sikka et al. (1980), and Tucker et al. (1985) was calculated by Layton et al. (1987) to be 2.17.

Twenty years after burial of RDX in outdoor soil plots at Los Alamos, NM, approximately 70% of the original concentration of 1 g/kg remained (DuBois and Baytos 1991). A half-life of 36 yr for RDX buried in soil was determined.

Biotransformation. The biodegradation of RDX has been studied under aerobic and anaerobic conditions. The data suggest that RDX is resistant to aerobic biodegradation as, with the exception of special conditions, it undergoes no or limited aerobic biodegradation when tested using a variety of inoculums, nutrients, and microbial sources (McCormick et al. 1981; Osmon and Klausmeier 1972; Spanggord et al. 1980b, 1983). Adaptation of microbes to RDX appears to be a necessary precondition for aerobic biodegradation to take place. Biotransformation takes place under anaerobic conditions.

Spanggord et al. (1980b) conducted a biotransformation screening test with Holston River water. RDX (10 ppm) was not degraded during 78 d of incubation. The addition of 30 ppm yeast extract did not significantly affect the disappearance. When Holston River sediment was added to the solutions, RDX remained unchanged for 20 d and was then reduced to 4 ppm between d 20 and d 36, followed by no change during the next 2 wk. Sikka et al. (1980) also found that, following a lag period, mineralization occurred in Holston River water to which sediment had been added. In another trial experiment, no degradation of RDX was observed during a 90-d aerobic experiment in which RDX was placed in lagoon water alone, with added yeast extract, or with 1% of bottom sediment (Spanggord et al. 1983). Concentrations also remained unchanged when cultures were inoculated with aerobic activated sludge and incubated aerobically (McCormick et al. 1981).

The degradation of pink water compounds in soil columns was studied by Greene et al. (1984). A pink water solution containing 30 mg/L RDX was continuously applied to a series of soil columns at different flow rates, with and without carbon supplementation. The soil columns were filled with garden soil and inoculated with microorganisms from activated sludge, anaerobic sludge digest, and garden soil. There was a significant decrease in RDX recovery in the leachate of the columns with the slowest and fastest flow rates, indicating microbial activity. Some biodegradation took place as indicated by the identification of hexahydro-1-nitroso-3,5-dinitro-1,3,5-triazine (MNX) and hexahydro-1,3-dinitroso-5-nitro-1,3,5-triazine (DNX) in the leachates. When Palouse soil was amended with ^{14}C-labeled-RDX and sampled 60 d later, only RDX was recoverable; $^{14}CO_2$ accounted for only 0.31% of the amended radiolabel (Harvey et al. 1991). Cataldo et al. (1993) treated soils containing different amounts of organic matter with 60 ppm of RDX. After 60 d, >95% of the RDX was extractable as the parent compound with <2% being nonextractable. No significant transformation products were identified.

In specific inoculum testing of aqueous solutions, a mixed population of purple photosynthetic bacteria of the genera *Chromatium, Rhodospirillum*, and *Rhodopseudomonas*, and possibly others, degraded 97% of a RDX solution in 5 d (Soli 1973). In specific genera testing, 60% of the biodegradation activity was attributable to *Chromatium* sp. alone. These cultures did not release oxygen, and it was hypothesized that degradation was caused by electron transfer rather than metabolism.

Data are available indicating that biodegradation of RDX occurs primarily under anaerobic conditions. RDX (30 mg/L) when added to Holston River water along with yeast extract and incubated in anaerobic flasks was reduced to <0.1 mg/L within 10 d (Spanggord et al. 1980b). No transformation was observed in flasks without added yeast extract. Additional studies determined that yeast extract serves as a cometabolic substrate. Using lagoon water alone or lagoon water with sediment from the Louisiana AAP, anaerobic incubation of RDX with yeast extract resulted in transformation only after an 80- to 90-d acclimation period (Spanggord et al. 1983).

McCormick et al. (1981, 1984) incubated RDX (50 or 100 μg/mL) anaerobically in nutrient broth cultures inoculated with anaerobic sewage sludge. RDX disappeared rapidly, and biodegradation was complete after 4 d. The disappearance of RDX was accompanied by the appearance of several products identified as the mono-, di-, and trinitroso RDX derivatives formed by sequential reductions of the nitro groups to nitroso groups. The metabolites formed were MNX, DNX, hexahydro-1,3,5-trinitroso-1,3,5-triazine (TNX), hydrazine, 1,1-dimethyl-hydrazine, 1,2-dimethyl-hydrazine, formaldehyde, and methanol.

B. Aquatic Toxicology

1. Acute Effects: Invertebrates and Fish. Results of acute tests conducted under static and flow-through conditions with four species of invertebrates and four species of fish are reported in Table 41. These tests were conducted under the USEPA (1975) guidelines available at that time. EC_{50} values, based on immobilization, were >100 mg/L for invertebrates under static conditions, whereas values were lower under flow-through conditions, >15 mg/L for the water flea,

Table 41. Acute toxicity of RDX to aquatic invertebrates and fish.

Test species	Test duration (test type)[a]	EC_{50} or LC_{50} (mg/L)[b,c]	Reference
Daphnia magna	48-hr (s)	>100	Bentley et al. 1977a
(water flea)	48-hr (ft)	>15	
Gammarus fasciatus	48-hr (s)	>100	Bentley et al. 1977a
(scud)			
Asellus militaris	48-hr (s)	>100	Bentley et al. 1977a
(sowbug)			
Chironomus tentans	48-hr (s)	>100	Bentley et al. 1977a
(midge)	48-hr (ft)	>15	
Ictalurus punctatus	96-hr (s)	4.1	Bentley et al. 1977a
(channel catfish)	96-hr (ft)	13	
Lepomis macrochirus	96-hr (s)	6.0	Bentley et al. 1977a
(bluegill sunfish)	96-hr (ft)	7.6	
Lepomis macrochirus	96-hr (s)	3.6–8.4[d]	Bentley et al. 1977a
(bluegill sunfish)			
Pimephales promelas	96-hr (s)	5.8	Bentley et al. 1977a
(fathead minnow)	96-hr (ft)	6.6	
Pimephales promelas			
(fathead minnow)			
Eggs	96-hr (s)	>100	Bentley et al. 1977a
1-hr post-hatch fry	96-hr (s)	43	Bentley et al. 1977a
7-d post-hatch fry	96-hr (s)	3.8	Bentley et al. 1977a
30-d post-hatch fry	96-hr (s)	16	Bentley et al. 1977a
60-d post-hatch fry	96-hr (s)	11	Bentley et al. 1977a
Pimephales promelas	96-hr (s)	4.5	Liu et al. 1983b
(fathead minnow)			
Oncorhynchus mykiss	96-hr (s)	6.4	Bentley et al. 1977a
(rainbow trout)			

[a]Test type: s, static; ft, flow-through.
[b]EC_{50} values are for immobilization of invertebrates; LC_{50} values are for fish mortality.
[c]Concentrations are nominal.
[d]Tested at temperatures of 15°–20 °C, pH, of 6.0–8.0; water hardness, 35–250 mg/L $CaCO_3$.

Daphnia magna, and the midge, *Chironomus tentans*. For fish, LC_{50} values ranged from 4.1 to 13 mg/L, with values slightly higher under flow-through conditions. Under static test conditions, the channel catfish (*Ictalurus punctatus*) was the most sensitive species, with an LC_{50} value of 4.1 mg/L. Bentley et al. (1977a) also explored the variation of acute toxicity with water quality parameters including temperature, pH, and hardness on the LC_{50} for the bluegill sunfish (*Lepomis macrochirus*) and found essentially no difference. The lowest 96-hr LC_{50} for bluegill sunfish was 3.6 mg/L at 20 °C, pH 6.0, and water hardness 35 mg/L $CaCO_3$.

Several different life stages of the fathead minnow (*Pimephales promelas*) were tested for 96 hr in static tests (Bentley et al. 1977a). The egg and 1-d posthatch fry were the least sensitive stages tested, and the 7-d posthatch fry was the most sensitive stage tested with an LC_{50} of 3.8 mg/L.

2. *Chronic Effects: Invertebrates and Fish.* Bentley et al. (1977a) conducted chronic tests with two species of invertebrates and two species of fish (Table 42). The protocol included both concurrent controls and vehicle controls (dimethylsulfoxide used to dissolve the RDX). RDX had almost no effect on survival of the parent generation of *D. magna* when tested at mean measured concentrations of 1.4, 2.2, 4.8, 9.5, or 20 mg/L, but the mean number of young produced per parthenogenic female exposed to concentrations of 4.8, 9.5, or 20 mg/L was significantly reduced between 7 and 14 d. At the lower concentrations from d 7 to 14 and at all concentrations from d 14 to 21 and d 28 to 42 (second generation), the daphnid production rate was comparable to controls. However, problems with maintaining RDX concentrations during d 7–14 and borderline acceptable survival of controls (78%–82%) render the data from this part of the study of limited value.

The midge *C. tentans* was tested at concentrations of 1.3, 2.2, 4.0, 10, and 21 mg/L. RDX had no statistically measurable effects on the larvae, pupae, adult survival, or adult emergence in the first generation. However, the average number of eggs produced per adult (treated throughout the larval and pupal stage) was greatly reduced compared with controls, with no eggs produced at the 10 mg/L concentration and no fertile eggs at 1.3 and 4.0 mg/L. During a second-generation exposure, eggs from the control containers were substituted in the 1.3, 4.0, and 10 mg/L concentrations. No significant differences from controls were observed in percent survival of pupae at all levels and in survival of adults at 2.2, 10, and 21 mg/L exposures. A significant reduction in second-generation adult emergence at 2.2 mg/L, in larval survival at all concentrations, and in adult survival at 1.3, and 4.0 mg/L was not conclusively related to RDX exposures because second-generation studies at 1.3, 4.0, and 10 mg/L were initiated with control eggs. Furthermore, no effects occurred at these concentrations in the first generation.

Thirty-day tests were conducted with early life stages of fish by Bentley et al. (1977a). Channel catfish were tested at 0.11, 0.30, 0.71, 1.2, and 2.3 mg/L and fathead minnows at 0.26, 0.76, 1.2, 3.0, and 5.8 mg/L. For channel catfish,

Table 42. Chronic toxicity of RDX to aquatic invertebrates and fish.[a,b]

Species	Stage/age	Parameter measured	Parameter response	Concentration (mg/L)	Reference
Daphnia magna (water flea)	Neonates through generation two	21-d survival (generation one)	No effect	1.4–20	Bentley et al. 1977a
		Reproduction	Significant decrease, d 7-14 of first generation	4.8–20	
Chironomus tentans (midge)	Neonates through generation two	23-d survival/emergence (generation one)	No effect	1.3–21	Bentley et al. 1977a
		Reproduction (21 d)	Lower larval survival	1.3–21	
			Lower egg production	1.3–10	
Ictalurus punctatus (channel catfish)	Eggs, fry	Mean hatch	No effect	0.11–2.3	Bentley et al. 1977a
		30-d fry survival	No effect	0.11–0.71[c]	
		30-d fry length	No effect	0.11–2.3	
Pimephales promelas (fathead minnow)	Eggs, fry	Mean hatch	No effect	0.26–5.8	Bentley et al. 1977a
		30-d fry survival	No effect	0.26–5.8	
		30-d fry length	Reduced	5.8	
Pimephales promelas (fathead minnow)	Eggs, fry, adults	Percent hatch	No effect	0.29–6.3	Bentley et al. 1977a
		30-d fry survival, length	No effect	0.29–4.3	
		60-d fry survival	Reduced	4.9	
		30, 60-d fry survival	No effect	3.0	
			Reduced	6.3	
		Spawning of 1st generation	No effect	6.3	
		2nd-generation parameters	No effect	6.3	

[a]Flow-through conditions.
[b]Mean measured concentrations except 30-d egg/fry tests, which utilized measured concentration for 3/5 treatments.
[c]Equipment failure at higher concentrations.

the tested concentrations had no effect on hatching or fry length at 30 d; poor fry survival at 30 d at the two highest concentrations was attributed to equipment malfunction. For fathead minnows, mean hatch and 30-d survival were not affected by concentrations of 0.26–5.8 mg/L; fry length at 30 d was significantly reduced at 5.8 mg/L.

Chronic tests with fathead minnows also resulted in minimal effects (Bentley et al. 1977a). Fish were tested at 0.29, 0.64, 1.1, 2.7, and 4.9 mg/L for 140 d and at 0.43, 0.78, 1.5, 3.0, and 6.3 mg/L for 240 d. In the first test, concentrations as high as 4.9 mg/L had no significant effect on the percentage of fry hatching or on fry survival and total length during the initial 30 d of exposure. After 60 d, survival of fry exposed to 4.9 mg/L RDX was statistically significantly lower than survival of controls or lower treatment groups, but total length at 60 d was unaffected. In the second test, percentage hatchability was not affected at any concentration, but after 30 and 60 d, survival of fry exposed to 6.3 mg/L was significantly lower than control survival. Fry length was unaffected by treatment after 30 and 60 d. After 240 d, total lengths and wet weights of mature male and female fathead minnows were not affected by exposure to concentrations of RDX as high as 6.3 mg/L. The spawning activity, percentage hatchability, and number of eggs were essentially unchanged by the presence of RDX in concentrations ≤6.3 mg/L. The second generation of fathead minnows also exhibited no measurable effects after 30 d of exposure to RDX at concentrations ranging up to 6.3 mg/L. Bentley determined a NOEC and LOEC in these studies of 3.0 and 6.3 mg/L, respectively.

3. Plants. Bentley et al. (1977a) examined the toxicity of RDX to two species of blue-green algae, *Microcystis aeruginosa* and *Anabaena flos-aquae*, the green alga *Selenastrum capricornutum*, and a species of diatom, *Navicula pelliculosa*, in static tests at nominal concentrations of 0.32–32 mg/L (Table 43). Parameters measured were chlorophyll a content and cell numbers. Although Bentley et al. (1977a) reported 96-hr EC_{50} values >32 mg/L, all four species showed significant decreases in chlorophyll *a* content and cell density, especially at the higher concentrations. Sullivan et al. (1979) reexamined the test with *S. capricornutum* and reported a significant reduction of chlorophyll *a* content at all concentrations. However, although statistically significant, the biological significance of a 3% decrease (at a concentration of 0.32 mg/L) is questionable. Therefore, effect levels reported in Table 43 are ≥10% changes in the reported parameters. *Selenastrum capricornutum* was the most sensitive species with a 16% reduction in chlorophyll *a* content at 1.0 mg/L and a 17% decrease in cell density at 3.2 mg/L.

4. Metabolism and Bioconcentration. Studies on the metabolism of RDX by aquatic organisms were not located in the available literature. Studies with radiolabeled RDX show that RDX does not bioconcentrate in aquatic organisms. Bentley et al. (1977a) calculated the bioconcentration of RDX in bluegill sunfish, channel catfish, and fathead minnows exposed continuously for 28 d to

Table 43. Toxicity of RDX to algae.[a,b]

Test species	Test duration	Effect	Concentration (mg/L)	Reference
Anabaena flos-aquae (blue-green alga)	96-hr	Decrease in cell density (17%)	32	Bentley et al. 1977a
		Decrease in chlorophyll *a* content (17%)	32	
Microcystis aeruginosa (blue-green alga)	96-hr	Decrease in cell density (18%)	32	Bentley et al. 1977a
		Decrease in chlorophyll *a* content (11%)	10	
Navicula pelliculosa (diatom)	96-hr	Decrease in cell density (17%)	32	Bentley et al. 1977a
		Decrease in chlorophyll *a* content (23%)	32	
Selenastrum capricornutum (green alga)	96-hr	Decrease in cell density (17%)	3.2	Bentley et al. 1977a
		Decrease in chlorophyll *a* content (16%)	1.0	

[a]All tests were conducted under static conditions.
[b]Nominal concentrations.

^{14}C-labeled RDX in an intermittent-flow system (Table 44). Mean measured concentrations were 0.014 and 1.0 mg/L. Tissues were wet-weighed, air-dried, and combusted, with the resulting $^{14}CO_2$ trapped. Bioconcentration of ^{14}C-labeled RDX residues in the edible tissues and viscera of these species reached a steady state after 14 d of exposure. Uptake in viscera was about two to three times that found in muscle, and the ranges of BCFs in muscle and viscera for all fish were 2.9–5.9 and 3.3–11, respectively. The mean BCF values for muscle at the two tested concentrations for channel catfish, bluegill sunfish, and fathead minnows were 3.4, 4.1, and 5.0, respectively. It was also noted that elimination of all RDX from the tissues at the lower concentration (0.014 mg/L) occurred after 14 d in the bluegill sunfish and catfish, but did not occur totally in the fathead minnow. At the higher dose, minnows and catfish eliminated 70%–87% of the accumulated RDX, but virtually no elimination occurred from either muscle or viscera in the bluegill sunfish.

Table 44. Bioconcentration factors (BCF) for RDX in aquatic species.

Species	BCF	Comments	Reference
Selenastrum capricornutum (green alga)	123	96-hr static test, intact organism	Liu et al. 1983b
Lumbriculus variegatus (oligochaete)	3.0	96-hr static test, intact organism	Liu et al. 1983b
Daphnia magna (water flea)	1.6	96-hr static test, intact organism	Liu et al. 1983b
Ictalurus punctatus (channel catfish)		28-d, intermittent flow:	Bentley et al. 1977a
	4.0[a], 2.9[b]	Muscle	
	5[a], 3.3[b]	Viscera	
Lepomis macrochirus		96-hr static test:	Liu et al. 1983b
(bluegill sunfish)	1.9	Muscle	
	3.1	Viscera	
Lepomis macrochirus (bluegill sunfish)		28-d, intermittent flow:	Bentley et al. 1977a
	4.7[a], 3.5[b]	Muscle	
	9[a], 6[b]	Viscera	
Pimephales promelas (fathead minnow)		28-d, intermittent flow:	Bentley et al. 1977a
	5.9[a], 4.0[b]	Muscle	
	11[a], 8.8[b]	Viscera	
Fish	2	Calculated from log K_{ow}	Layton et al. 1987
Fish	1.5	Calculated from log K_{ow}	Burrows et al. 1989

[a]Test water mean measured concentration of 0.014 mg/L.
[b]Test water mean measured concentration of 1.0 mg/L.

Bioconcentration was also studied by Liu et al. (1983b) using four aquatic species. Tests were performed using a 96-hr static exposure to ^{14}C-labeled RDX at a concentration of 0.3 mg/L. Bioconcentration factors were estimated using the average amount of radioactivity recovered (disintegrations/min/g) in tissue and water. The reported 4-d BCFs are listed in Table 44. The 96-hr BCF calculated for bluegill sunfish muscle (1.9) and viscera (3.1) are somewhat lower than the 72-hr BCF reported by Bentley et al. (1977a), i.e., 3.5 and 9.5, respectively (data not shown), when tested at a concentration of 0.014 mg/L, and 4.3 and 13, respectively, when tested at a concentration of 1.0 mg/L. Based on log K_{ow} values of 0.81–0.87, Burrows et al. (1989) calculated a BCF of 1.5 for fish.

C. Aquatic Criteria and Screening Benchmarks

1. Aquatic Organisms. With one exception, the acute toxicity studies were all conducted by the same investigator. In the case of *P. promelas*, the two static 96-hr toxicity tests conducted by separate investigators yielded similar results. The chronic studies, unfortunately, suffered from reported equipment failures and, at times, inconsistent results. Nevertheless, because most of the available data were consistent, they were used to estimate aquatic criteria. These estimates can be used as screening benchmarks for potential adverse effects on aquatic organisms.

Data were insufficient for calculation of acute and chronic WQC according to USEPA guidelines (Stephan et al. 1985). Data are available for only six of the required eight families: a planktonic crustacean (*D. magna*), the family Salmonidae (*O. mykiss*), a second family in the class Osteichthyes (*L. macrochirus*), a third family in the phylum Chordata (*P. promelas* or *I. punctatus*), an insect (*C. tentans*), and a benthic amphipod (*Gammarus fasciatus*). Therefore, Tier II or secondary acute and chronic values were calculated according to the USEPA guidance for the Great Lakes System (USEPA 1993a).

Although flow-through tests with measured concentrations are recommended by Stephan et al. (1985), the static tests were available for all fish species and had lower, more conservative values. Thus, they were included in the calculations. Values for invertebrates were considerably higher than those for fish. Values used to calculate the GMAV for the fish species (genera) were 4.1 and 13 mg/L for *I. punctatus*, 6.0 and 7.6 mg/L for *L. macrochirus*, 5.8, 6.6, and 4.5 mg/L for *P. promelas*, and 6.4 mg/L for *O. mykiss*. Other values for these species listed in Table 41 were not used, either because the fish were not of the appropriate age or size or because the tests were not conducted under the appropriate physicochemical conditions. The lowest GMAV was 5.56 mg/L for *P. promelas*. USEPA (1993a) lists a SAF of 4.0 for use in Tier II calculations when six satisfied data requirements for Tier I calculations are available.

Therefore

$$\text{SAV} = \frac{5.56}{4.0} = 1.39 \text{ mg/L}$$

Table 45. Aquatic acute/chronic ratios (ACR) for RDX.

Species	Acute values (mg/L)	Chronic values (mg/L)	Acute/chronic ratios (ACR)
Rainbow trout	5.56	4.35	1.128
Default ACR[a]			18
Default ACR			18

[a]In the absence of three ACRs, two default ACRs of 18 were substituted (USEPA 1993a).

The SMC is one-half of the SAV or 0.6955 mg/L.

One of three required experimentally determined ACR were available according to guidelines for Tier I (chronic tests on a sensitive species and an invertebrate were not acceptable). Therefore, two assumed ACRs of 18 were added. Although the tests conducted by Bentley et al. (1977a) suffered from equipment malfunctions, for fathead minnows, the data appear to be consistent for the two tests reported in Table 42, and a NOEC (3.0 mg/L) and LOEC (6.3 mg/L) (geometric mean, 4.35 mg/L) were determined in a life cycle test. In addition, these are the only available data. For *D. magna*, the acute/chronic ratio of >15 mg/L/2.6 mg/L = >5.8 mg/L is not acceptable and will be replaced by the default ACR of 18. The SACR is the geometric mean of the three ACRs listed in Table 45, i.e., 7.45 mg/L.

The SCV is the SAV divided by the SACR:

$$\mathrm{SCV} = \frac{\mathrm{SAV}}{\mathrm{SACR}} = \frac{1.39}{7.455} = 0.186\ \mathrm{mg/L}$$

Screening criteria/benchmarks as well as lowest chronic values are listed in Table 46.

Table 46. Water quality criteria/screening benchmarks for RDX.

Criterion/benchmark	Value
Acute water quality criterion	Insufficient data
Chronic water quality criterion	Insufficient data
Secondary acute value	1.4 mg/L[a]
Secondary maximum concentration	0.7 mg/L
Secondary chronic value	0.19 mg/L
Secondary continuous concentration	0.19 mg/L
Lowest chronic value, fish	4.9 mg/L
Lowest chronic value, daphnids	4.8 mg/L
Final plant value	3.2 mg/L[a]
Sediment quality benchmark (SQB_{oc})	1.3 mg/kg_{oc}[b]

[a]Estimated; based on nominal concentrations.
[b]mg chemical/kg organic carbon in sediment.

A FPV can be estimated from the study of Bentley et al. (1977a). Although concentrations were nominal rather than measured, the test by in which *S. capricornutum* showed a decrease in cell density at a concentration of 3.2 mg/L can be used as a screening value for the FPV. In other tests conducted by Bentley et al. (1977a) in which concentrations were measured, nominal values were close to measured values.

Before publication of the methodology for Tier II values, Bentley et al. (1977a) used an application factor of 0.1 for the bluegill sunfish data to derive a WQC of 0.35 mg/L, Sullivan et al. (1979) recommended a 24-hr average concentration of 0.3 mg/L to adequately protect aquatic life, and Etnier (1986) estimated a FAV of 5.2 mg/L.

2. Sediment-Associated Organisms. No data on the toxicity of spiked sediments or the interstitial water from spiked sediments to sediment-associated organisms were located for this chemical. One study was found that tested the toxicity of RDX to benthic organisms: the midge *C. tentans*, the isopod *Asellus militaris*, and the amphipod *G. fasciatus* (Bentley et al. 1977a); however, the tests were conducted in 250-ml beakers without sediment and cannot be used to derive a SQB. From these tests, it is known that a sediment pore-water concentration of 15 mg/L would not be acutely toxic to these organisms (see Table 41).

The EqP approach can be used to estimated a SQB. USEPA (1993b) recommends using a reliably measured octanol–water partition coefficient, K_{ow}, to estimate the K_{oc}. Because K_{ow} values are available, a K_{oc} can be calculated and a SQB_{oc} can be determined. Log K_{ow} values of 0.87, 0.86, and 0.81 were cited in Banerjee et al. (1980) and Burrows et al. (1989); the geometric mean of these values is 0.846. Using the relationship between K_{ow} and K_{oc}, the log K_{oc} is 0.832 and the K_{oc} is 6.79. Then, the organic carbon-normalized SQB_{oc}, i.e., mg chemical/kg organic carbon (mg/kg_{oc}) is

$$SQB_{oc} = K_{oc} \times SCV$$
$$SQB_{oc} = 6.79 \text{ L/kg} \times 0.186 \text{ mg/L} = 1.27 \text{ mg/kg}_{oc}$$

If an organic carbon content in the soil of 1% is assumed, the $SQB_{1\%}$ is calculated as

$$SQB_{1\%} = 0.01 \times 6.79 \text{ L/kg} \times 0.194 \text{ mg/L} = 0.01 \text{ mg/kg}$$

D. Terrestrial Toxicology

1. Mammals. No subchronic or chronic studies on the toxicity of RDX utilizing mammalian wildlife species were located, but three chronic toxicity studies and one combined teratogenicity/reproductive study on RDX using laboratory animals were found. RDX is known to have an effect on the central nervous system, with symptoms of convulsions and hyperactivity in both humans and laboratory animals (Ryon et al. 1984; Layton et al. 1987; USEPA 1989). In some of the following studies, these symptoms were observed in rats and mice only at the higher doses. In addition to neurotoxicity, effects on the blood, kidney, liver, eyes, and testes have been observed. Most evidence indicates that RDX is not teratogenic or carcinogenic. Studies are summarized in Table 47. The LOAELs listed in the table are for population-related effects (i.e., survival and reproductive parameters); for other endpoints (i.e., body or organ weight changes), lower LOAELS and NOAELs may have been reported in these studies.

In the first study, Hart (1976) administered RDX to groups of 100 male and 100 female Sprague-Dawley rats at doses of 0, 1.0, 3.1, or 10 mg/kg/d for 104 wk. Observations were weekly or biweekly and sacrifices were performed at 52 and 104 weeks. At termination, there were no statistically significant differences between control and treatment groups in survival, food consumption, body or organ weight, or histopathological findings. Fluctuations in clinical chemistry parameters and hematocytology occurred but were not dose related.

In the second study, groups of 75 male and 75 female Fischer-344 rats were administered RDX (containing 3%–10% HMX) in the diet at doses of 0, 0.3, 1.5, 8.0, or 40 mg/kg/d for 24 mon (Levine et al. 1983). The incidence of mortality in the high-dose group (68% of the males and 36% of the females) was increased throughout the study and was frequently preceded by tremors and convulsions. Other treatment-related findings in this group were anemia with secondary splenic lesions, hypoglycemia, reduced body weight gains, hepatotoxicity, and possible CNS involvement. In this dose group, cataracts were observed in females and urogenital lesions were present in males. Rats of both sexes administered 8 mg/kg/d had body weights that were approximately 5% lower than those of controls and increased kidney weights. Females had hepatomegaly and decreased total serum protein levels, and males had an increase in hemosiderin-like pigment in the spleen and suppurative inflammation of the prostrate. The latter effects were also observed in males receiving 1.5 mg/kg/d. The latter effect may have been caused by a bacterial infection because bladder distension and cystitis were also noted. No adverse effects were observed in rats administered 0.3 mg/kg/d.

In the third study, RDX (containing 3%–10% HMX) was administered in the diet to groups of 85 male and 85 female $B6C3F_1$ mice at doses of 0, 1.5, 7.0, 35, or 100 mg/kg/d for 24 mon (Lish et al. 1984). The 100 mg/kg/d dose group was originally administered 175 mg/kg/d, but due to high mortality through test wk 10, the dose was lowered to 100 mg/kg/d beginning with test wk 11. Following additional early mortalities after lowering the high dose to 100 mg/kg/d, survival in the remaining animals was similar to controls. Males receiving the 175 mg/kg/d dose had a high incidence of fighting wounds and skin lesions, but incidences of these lesions were similar in all groups by the end of the test. Food consumption was not altered by RDX treatment, but final body weights were lower than controls in the high-dose group. Hematology parameters were not altered in any dose group. Elevated serum cholesterol levels were present in high-dose males and in females receiving 7 and 35 mg/kg/d. Cataracts were

Table 47. RDX toxicity data for mammalian species.

Species	Route	Exposure period	NOAEL (mg/kg/d)	LOAEL (mg/kg/d)	Effect/endpoint	Reference
Rat	Diet	2 yr	10[a]		No effects	Hart 1976
	Diet	2 yr	8[b]	40	Mortality	Levine et al. 1983
	Diet	13 wk/generation (2 generations)	16[c,d]	50[e]	Mortality	Cholakis et al. 1980
	Diet	13 wk/generation (2 generations)	5	16[d]	Reduced pup weight	Cholakis et al. 1980
	Gavage	Gestation d 6–19	2	20[f]	Toxic to dams and embryos	Cholakis et al. 1980
Mouse	Diet	2 yr	7	35	Testicular degeneration	Lish et al. 1984
Rabbit	Gavage	Gestation d 7–29	2	20[f]	Toxic to dams and embryos	Cholakis et al. 1980

[a]Maximum dose tested.

[b]Inflammation of prostate seen at 1.5 and 8.0 mg/kg/da; NOEL, 0.3 mg/kg/d.

[c]No adverse reproductive effects seen at this dose.

[d]Nominal dose; calculated doses were 13.8 (males) and 14.5 (females) mg/kg/d in the F_0 generation and 17.6 (males) and 18.8 (females) mg/kg/d in the F_1 generation.

[e]Nominal dose; calculated doses were 41.3 (males) and 43.6 (females) mg/kg/d in the F_0 generation and 61.3 (males) and 69.6 (females) mg/kg/d in the F_1 generation.

[f]No teratogenic effects seen at this dose.

present in high-dose males but not in females. Organ weights were altered in some dose groups: increased liver and kidney weight in both sexes receiving 100 mg/kg/d, increased kidney weight in males receiving 35 mg/kg/d, and increased heart weights in both sexes receiving 100 mg/kg/d. Histopathological examinations revealed no apparent lesions at the 12-mon sacrifice. By 24 mon the following statistically significantly elevated lesions were observed: testicular degeneration in males administered 35 and 100 mg/kg/d, liver carcinomas in females administered 35 and 100 mg/kg/d, and malignant lymphoma of the liver in males administered 35 mg/kg/d. When combined liver adenoma/carcinoma data were compared to both historical and concurrent control groups, the incidence was also statistically significant in females receiving 7 mg/kg/d.

The teratogenicity of RDX was studied in rats and rabbits (Cholakis et al. 1980). Rats were dosed by gavage on d 6 through 19 of gestation and rabbits were gavaged on d 7 through 29 of gestation with 0, 0.2, 2.0, or 20 mg/kg/d of RDX. Dams were monitored daily for toxic signs. On d 20 of gestation (rats) or d 30 of gestation (rabbits), animals were sacrificed and the uterus of each animal was examined for live fetuses and resorptions. Fetuses were examined grossly and microscopically for malformations.

A dose of 20 mg/kg/d was toxic to rat dams, resulting in mortality (6/25 animals) and reduced body weight, reduced feed consumption, neurotoxic signs, and reduced liver weights of the survivors. These signs were not noted in the other dose groups. The dose of 20 mg/kg/d was also embryotoxic. There were no effects on reproductive parameters (number of implants and viable fetuses) or increases in malformations of fetuses of surviving dams in the 20 mg/kg/d dose group or the lower dose groups. No treatment-related adverse effects were noted in pregnant rabbits. For rabbit fetuses, there were no statistically significantly increased incidences of resorptions or deaths or gross or microscopic anomalies at any dose level.

In the two-generation reproduction study, groups of 22 male and 22 female rats were fed diets designed to deliver doses of RDX of 0, 5, 16, or 50 mg/kg/d (Cholakis et al. 1980). After 13 wk the rats (F_0 generation) were mated. The resultant F_1 pups were weaned, and groups of 26 males and 26 females per dose group were administered respective diets for 13 wk. After this treatment males and females were mated as before and resulting litters (F_2 generation) were sacrificed and examined microscopically.

Effects were observed primarily in the group receiving the 50 mg/kg/d dose during and following the F_0 mating: deaths before mating, reduced fertility in males and females (not statistically significant), reduced number of litters (not statistically significant), and reduced viability of pups in the high-dose group (statistically significant). The body weights of pups in both the mid- and high-dose groups were reduced after 25 d, and only one litter in the high-dose group survived to d 21. Meaningful comparisons between the high-dose group and the other groups and controls following the F_1 mating are difficult to make because only four males and two females survived until wk 13 in the F_1 high-dose group compared with 26 mating pairs in the other groups. However, pup survival was obviously affected as the number of pups/litter on d 0 was statistically significantly reduced compared with the control group and no pups in this dose group survived more than a few days. Histopathological evaluations revealed a statistically significant increase in renal cortical cysts in females administered 16 mg/kg/d. Body weights of females were also reduced.

2. Birds. No subchronic or chronic studies on the toxicity of RDX to birds were located.

3. Plants. Simini et al. (1992) performed soil toxicity testing with the cucumber (*Cucumis sativus*) at RDX concentrations of 0, 50, 100, or 200 mg/kg of soil. Biomass of plants exposed to 100 and 200 mg/kg was significantly reduced. The results were reported in an abstract, and no further information was given. Simini et al. (1995) also evaluated the toxicity of Joliet AAP (IL) soil, which contained a mixture of munitions compounds, to radishes and cucumbers. RDX was found to be very weakly correlated with a decrease in plant height.

4. Soil Invertebrates. Simini et al. (1995) tested the toxicity of munitions-contaminated soil (contaminated in part with RDX) and calculated the correlation of RDX concentration with overall soil toxicity. Endpoints were survival rates and growth of earthworms (*Eisenia foetida*). RDX was very weakly correlated with these endpoints. RDX concentrations ranged from below the limit of detection in most samples (0.1 mg/kg) to 3574 mg/kg.

In a preliminary study, Phillips et al. (1993) tested the toxicity of RDX in artificial soil (1.4% organic matter) to the earthworm *E. foetida* over a 14-d period. Concentrations of RDX were 0, 50, 100, 200, 400, and 500 mg/kg. Although survival was 80%–100% for all concentrations, there was an increasing weight loss with increasing concentrations. Weight losses ranged from 11% at the 50 mg/kg concentration to 18% at the 500 mg/kg concentration.

5. Soil Heterotrophic Processes. No data on the effect of RDX on microorganisms in soil were located. In biodegradation studies, concentrations up to 50 mg/L were not toxic to microorganisms in anaerobic sewage sludge (McCormick et al. 1984).

6. Metabolism and Bioaccumulation.

Animals. RDX is slowly but extensively absorbed following ingestion. Based on a lack of physiological response following dermal application to dogs and rabbits, McNamara et al. (1974) concluded that RDX did not penetrate the skin. No inhalation data were located. Absorbed RDX is readily metabolized and excreted by mammals and does not accumulate in any tissue (Schneider et al. 1978). Following administration by gavage or in the drinking water to rats for 90 d, RDX was distributed throughout the body with no preferential uptake by any specific organ or tissue including the fat (Schneider et al. 1978). The pattern of distribution was similar in miniature swine administered a single oral dose (Schneider et al. 1977). Metabolism in the rat is extensive as indicated by excretion of 43% of a single ^{14}C-radiolabeled oral dose as $^{14}CO_2$ within 4 d (Schneider et al. 1977). Thirty-four percent was excreted in the urine and 3% in the feces, while 10% remained in the carcass. A similar pattern was observed following administration to rats for 13 wk: 27%–50.5% as $^{14}CO_2$, 22%–35% in the urine, and 4%–5% in the feces (Schneider et al. 1978). Urinary metabolites were not identified. Log K_{ow} values of 0.87 (Banerjee et al. 1980), 0.81, and 0.86 (Burrows et al. 1989) also indicate a low potential for bioaccumulation.

Biomonitoring studies have been conducted at Aberdeen Proving Ground, MD (USACHPPM 1994), and at several other AAPs (see Opresko 1995b for review). RDX was not present in tissues of terrestrial wildlife (deer and small mammals) at or above a detection limit of 0.1 mg/kg.

Plants. Data from several studies indicate that RDX can be taken up by plants. When grown in hydroponic nutrient solutions containing 10 ppm RDX, uptake and translocation to the stems and leaves occurred (Harvey et al. 1991). After 7 d, concentrations of RDX in leaves, stems, and roots were 97, 11, and 6 ppm respectively, indicating accumulation in the aerial parts. RDX remained sequestered within the plants, with no evolution of $^{14}CO_2$.

To investigate the transformation of RDX before plant uptake, transformations were studied in hydroponic culture solutions with and without plant roots (unspecified species) (Harvey et al. 1991). No significant transformations took place in the solution. When 21- to 26-d-old bush bean plants were grown in solutions containing 10 mg/L ^{14}C-labeled RDX, uptake and translocation of the parent compound to the leaves readily occurred, with recovery of primarily RDX. Some metabolism to unidentified polar metabolites took place by d 7. RDX remained sequestered within the plants, with no evolution of $^{14}CO_2$.

In a follow-up study, bean, wheat, or blando brome were planted in soils amended with RDX at a concentration of 10 ppm (mg/kg, air-dried weight) and after 60 d the plants were analyzed by HPLC and radioactivity for total uptake, tissue distribution, and transformations (Cataldo et al. 1993). Three soils were tested, Palouse (1.7% organic matter), Cinebar (7.2% organic matter), and Burbank (0.5% organic matter) (Cataldo et al. 1989). RDX was accumulated by and was mobile in the plants (Cataldo et al. 1993). For all three plants, highest concentrations were found in the leaves before seed formation; following seed formation, highest concentrations were present in the seeds of bush bean. Total uptake and tissue concentration were inversely proportional to soil organic matter. In roots and stems, 20% of the translocated material was present as the parent compound, while in leaves and seed tissues >50% was RDX, suggesting that RDX itself is mobile within the plant. The balance of the accumulated radioactivity was either unidentified polar metabolites (30%–50%) or associated with insoluble plant parts.

Simini et al. (1992) studied the uptake of RDX by cucumber plants grown in RDX-amended soil. Plants grown in 50, 100, and 200 mg RDX per kg soil had 464, 1320, and 1702 mg RDX, respectively, per kg leaf tissue. When plants are grown in soil with a low organic carbon content and containing 10 ppm RDX, the compound is taken up, with as much as 603 μg RDX equivalents (RDX plus metabolites)/g fresh weight accumulated in the seeds of bush beans grown in soil containing 0.5% organic carbon (Cataldo et al. 1993). Lower concentrations were present in the roots, stems, and leaves.

E. Terrestrial Criteria and Screening Benchmarks

1. Mammals. A chronic oral RfD for humans of 0.003 mg/kg/d was calculated by extrapolation and multiplication by uncertainty factors from the chronic oral study with rats by Levine et al. (1983) (USEPA 1993f). The endpoint in that study was inflammation of the prostate.

Three chronic toxicity studies and a two-generation reproduction study on the oral toxicity of RDX utilizing laboratory animals were available for calculating screening benchmarks for mammalian wildlife (Cholakis et al. 1980; Hart 1976; Levine et al. 1983; Lish et al. 1984). The first three studies involved chronic exposures, whereas the reproduction study by Cholakis et al. (1980) involved subchronic (13-wk) exposures during each of two generations. The study by Hart (1976) did not attain a LOAEL for population-related effects. In the other chronic study with rats (Levine et al. 1983), it is difficult to predict if the dose of 8 mg/kg/d, which resulted in inflammation of the prostate, is truly a NOAEL for reproductive effects or whether the effect was treatment related. Thus, the remaining highest NOAEL of 7 mg/kg/d in the study with mice (Lish et al. 1984) was chosen to derive screening benchmarks for wildlife. A clear NOAEL and LOAEL for reproductive effects (testicular degeneration) were established in this study.

Of the studies listed in Table 47, the Hart study provides the highest NOAEL (10 mg/kg/d) for a chronic exposure period. However, if this NOAEL is used to derive benchmarks for the other species listed on the table, the calculated benchmarks will be higher than the listed NOAELs; therefore, the more conservative approach would be to use the mouse chronic NOAEL of 7 mg/kg/d derived in the study by Lish et al. (1984). Weekly or biweekly mean body weights for the male mice in the 7 mg/kg/d group ranged from 24.1 g before the test began to 38.6 g after 53 wk and 37.8 g after 104 wk. The overall mean of weekly and biweekly means was calculated to be 35.9 g.

Screening benchmarks for food and water intake for selected wildlife species were derived from each study using the methodology described in the Introduction and illustrated in the previous section on TNT (Table 48). For the mink, the BCF for RDX was estimated from the mean log K_{ow} of 0.85. Confidence in the benchmarks is high because the database is extensive and covers various toxicological endpoints.

2. Birds. No subchronic or chronic studies were available for calculation of screening benchmarks for avian species.

Table 48. RDX screening benchmarks for selected mammalian wildlife species.

Wildlife species	Chronic NOAEL (mg/kg/d)	Screening benchmarks		
		Diet (mg/kg food)	Water (mg/L)	Piscivorous species (mg/L)[a]
Shorttail shrew	8.7	15	40	—
White-footed mouse	7.9	56	29	—
Meadow vole	6.7	58	49	—
Cottontail rabbit	2.9	15	30	—
Mink	3.0	22	31	6.7
Red fox	2.1	21	25	—
Whitetail deer	1.1	36	17	—

[a]Water concentration that incorporates dietary intake from both water and food consumption.

3. Plants. In the absence of criteria for terrestrial plants, LOEC values from the literature can be used to screen chemicals of potential concern for phytotoxicity (Will and Suter 1995a). Plants were able to grow for 7 d in hydroponic solutions containing 10 ppm RDX (Harvey et al. 1991), but these data are insufficient to calculate a screening benchmark. Two studies on the growth of plants in RDX-amended soil were available. In the first study, bean, wheat, and blando brome plants were grown in soil amended with 10 mg/kg RDX, but effects on growth were not reported (Cataldo et al. 1989). Simini et al. (1992) reported that RDX concentrations of 100 and 200 mg/kg significantly reduced the biomass of cucumber plants. From this study, a screening benchmark of 100 mg/kg was determined (Table 49). Confidence in the benchmark is low, as it is based on a single study involving a single species.

4. Soil Invertebrates. According to the method of Will and Suter (1995b), a 20% reduction in growth would be a threshold for significant effects. Although concentrations to 500 mg RDX/kg of artificial soil were not lethal to the earthworm *E. foetida*, weight losses occurred but were <20% at the highest concen-

Table 49. RDX screening benchmarks for terrestrial plants and invertebrates.

Screening benchmark	Value
Plants, solution	Insufficient data
Plants, soil	100 mg/kg[a]
Soil invertebrates	Insufficient data
Soil microbial processes	Insufficient data

[a]Based on a single study.

tration tested, 500 mg/kg (Phillips et al. 1993). In addition, RDX in other soil types may behave differently than in artificial soils, and additional studies are needed before a screening value for earthworms can be determined.

5. Soil Heterotrophic Processes. Insufficient data were located for calculation of a screening benchmark for soil heterotrophic processes. Studies with laboratory cultures were not considered because they utilized acclimated organisms.

VIII. Octahydro-1,3,5,7-Tetranitro-1,3,5,7-Tetrazocine

Octahydro-1,3,5,7-tetranitro-1,3,5,7-tetrazocine, commonly known as HMX (high melting explosive), is a colorless, crystalline solid that has 130% of the explosive power of TNT. HMX is used in nuclear devices to implode fissionable material and is a component of plastic explosives, solid fuel rocket propellants, and military munitions (McLellan et al. 1988). HMX is also formed during the manufacture of another explosive, hexahydro-1,3,5-trinitro-1,3,5-triazine or RDX. The properties of HMX and RDX are quite similar. Both compounds are manufactured by the Bachmann process in which hexamine is nitrated by nitric acid and ammonium nitrate in the presence of acetic anhydride and acetic acid (Lindner 1978). Chemical and physical properties are listed in Table 50.

Information on concentrations and fate in the environment, aquatic toxicity, and mammalian toxicity was located. Data were sufficient to derive screening benchmarks for aquatic (fish, invertebrates, and sediment-associated organisms) and terrestrial mammalian species. Data were insufficient for derivation of screening benchmarks for avian species or terrestrial plants, invertebrates, or microorganisms.

A. Environmental Fate

1. Sources and Occurrences. HMX is currently produced only at the Holston AAP in Kingsport, TN. More then 30 million pounds of HMX is believed to be produced each year (ATSDR 1994). Further processing may occur at facilities where LAP operations are performed by the U.S. military (Ryon et al. 1984). Environmental releases may also occur as a result of treatment and disposal of HMX- and RDX-containing wastes (i.e., RDX contains about 5% HMX [Kitchens et al. 1978]). HMX has been detected at several sites on the Superfund National Priorities List (ATSDR 1994).

Air. HMX may be atmospherically released by adsorption to airborne particulates released during munitions manufacturing (McLellan et al. 1988). However, this is not expected to be a major pathway for HMX pollution. Estimates of HMX release to the air are unavailable (ATSDR 1994).

Surface Water and Groundwater. Most HMX releases to the environment originate in wastewater from facilities that manufacture, process, or use HMX

Table 50. Chemical and physical properties of octahydro-1,3,5,7-tetranitro-1,3,5,7-tetrazocine (HMX).

Synonyms	Cyclotetramethylenetetranitramine octagen Tetramethylenetetranitramine	HSDB 1995f
CAS number	2691-41-0	HSDB 1995f
Molecular weight	296.20	HSDB 1995f
Physical state	Colorless, crystalline solid; four polymorphic forms: α, β, γ, δ	Yinon 1990
Chemical formula	$C_4H_8N_8O_8$	Yinon 1990
Structure	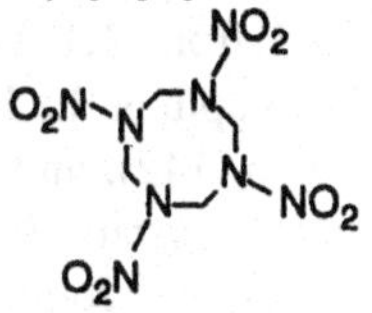	ATSDR 1994
Water solubility (20 °C)	6.6 mg/L	McLellan et al. 1988
Specific gravity	No data	
Melting point	286 °C	Lindner 1978
Boiling point	No data	
Vapor pressure	3.3×10^{-14} mm Hg	Burrows et al. 1989
Partition coefficients		
Log K_{ow}	0.26	Burrows et al. 1989
	0.13	Tucker et al. 1985
	0.06	Jenkins 1989
Log K_{oc}	2.8	Spanggord et al. 1982b
Henry's law constant (25 °C)	2.6×10^{-15} atm-m^3/mole	Burrows et al. 1989

(ATSDR 1994). HMX released to soil has been shown to partition to groundwater (Gregory and Elliott 1987).

Stidham (1979) reported that the Holston AAP was releasing 45 lb of HMX/d in wastewater to the Holston River, with effluent concentrations ranging from 0.09 to 3.36 mg/L. Levels as high as 67 µg/L occurred 1 mile downstream of the last plant effluent as a result of incomplete mixing of the waste stream with river water. Kitchens et al. (1978) estimated that at full mobilization at least 123 lb of HMX would be released per day. In this study, sampling at various points in the river and Holston AAP waste streams indicated concentrations of HMX between 0.01 (river) and 2.6 (wastewater) mg/L. Stillwell et al. (1977) also sampled various waste streams from the Holston AAP and found that untreated waste streams had HMX levels from 0.14 to 5.4 mg/L (except for one sample, which was below the detection limit of 0.1 mg/L), while treated waste streams had concentrations that ranged from <0.1 to 2.07 mg/L.

As part of a remedial investigation of the Louisiana AAP, Gregory and Elliott (1987) measured HMX levels in groundwater from four areas. Around the Area P leaching pits, HMX was detected in samples from 14 of 32 monitoring wells, with concentrations ranging from 3.12 to 4200 μg/L. At Burning Ground #5, levels ranged from 1.23 to 145 μg/L for 8 of 9 wells. At Inert Landfill Area #3, an HMX level of 2.28 μg/L was noted in 1 well of 13. At Inert Landfill Area #8, HMX concentrations of 8.3–157 μg/L were detected in samples from 6 of 19 wells.

Soils. Soil contamination occurs as a result of spills at Army facilities and disposal of HMX-contaminated wastes in landfills. Over a 10-yr period, Walsh and Jenkins (1992) collected dried soil samples from 17 Army sites, and analyzed the samples for a number of chemicals relating to munitions use and production. Levels of HMX in excess of the detection limit (1.1 ppm) were found at 5 of these sites; levels ranged from 1.1–115 ppm at the Nebraska Ordinance Works (median, 7.7), 2.5–5700 ppm at the Iowa AAP (median, 28.7), 56–2435 ppm at Hawthorne AAP (median, 142), and 3.7–86 ppm at the Milan AAP (median, 53), to 68–258 ppm at the Louisiana AAP (median 163). As part of a study designed to evaluate the toxicity of Joliet AAP soils, Simini et al. (1995) determined HMX concentrations in soil taken from a LAP area. Levels of HMX ranged from less than the limit of detection to 3055 ppm and exceeded the detection limit in 13 of 40 samples. Spiker et al. (1992) and Funk et al. (1993) reported that a soil sample from an Army munitions depot near Umatilla, OR, contained 300 ppm of HMX. Four soil samples taken from the Naval Surface Warfare Center contained concentrations ranging from 0.10 to 12.2 mg/kg (dry weight) (Grant et al. 1995).

2. *Transport and Transformation Processes.*

Abiotic Processes. HMX has a vapor pressure of 3.3×10^{-14} mm Hg and a Henry's law constant of 2.6×10^{-15} atm-m^3/mole at 25 °C (Burrows et al. 1989); thus, it will not partition to the air in the vapor phase. However, HMX may be atmospherically transported via adsorption to airborne particulates, especially those from munitions manufacture (ATSDR 1994). Photolysis will limit the persistence of airborne HMX as it does in the aquatic environment. However, no data are available on the fate of airborne HMX.

Direct photolysis is the primary pathway for degradation of HMX in the aquatic environment. Hydrolysis and oxidation of HMX are negligible (McLellan et al. 1988). Spanggord et al. (1982b) determined photolytic half-lives for HMX of 4–5 d in surface water from the Holston River. In a more detailed follow-up study, photolytic half-lives for HMX were determined to be 1.4, 1.7, and 70 d in pure water, Holston River water, and Louisiana AAP lagoon water, respectively (Spanggord et al. 1983). The rate of photolysis was apparently related to the amount of UV-absorbing material in the water; it was not affected by any cosubstrates present. Measured stable end products of photolysis were nitrate, nitrite, and formaldehyde. Dinitrogen oxide, ammonia, and formaldehyde were proposed as additional end products. Modeling studies on the fate of HMX in the Holston River and Louisiana AAP lagoon under their natural conditions (including such factors as water flow, depth, and seasonal changes) indicated that the half-life of HMX would be 17 d in the Holston River and 7900 d in the Louisiana AAP lagoon (Spanggord et al. 1983). HMX was predicted to persist in the Holston River for distances greater than 20 km downstream.

Volatilization, adsorption to suspended material, and biosorption are not expected to significantly remove HMX from the water column (Spanggord et al. 1982b). Volatilization half-lives of 1000–3000 d have been estimated. Partitioning of HMX to Holston River sediment was studied, and upper limit values of 8.7 for the sediment–water partition coefficient (K_p) and 670 for the organic carbon–water partition coefficient (K_{oc}) were determined. These values indicate a low to moderate affinity of HMX for soil or suspended material; thus, HMX dissolved in the water will not be removed though adsorption to sediment. For the same reason, dissolved HMX will readily migrate to groundwater from soil-contaminated areas. However, the low solubility of HMX (6.6 mg/L at 20 °C; McLellan et al. 1988) limits the partition to, and thus the migration of, HMX in groundwater.

Migration of HMX is greatest in coarse, loamy soil (Ryon et al. 1984). Direct photolysis will not occur in soil, and the chemical should be persistent in the soil environment (Ryon et al. 1984). This was confirmed by DuBois and Baytos (1991), who buried HMX in soil exposed to the environment at Los Alamos, NM, and calculated a half-life of 39 yr.

Spanggord et al. (1982b) calculated an upper-limit partition coefficient (K_p) for HMX on Holston River sediment of ≤8.7 L/kg. The K_p was calculated from a linear least-squares plot of HMX water concentration versus HMX sediment concentration (dry weight) from water–sediment samples prepared from HMX-contaminated water and Holston River sediment. The organic content of the sediment was determined to be 1.3%, from which a K_{oc} value of ≤670 L/kg was calculated. These values indicate that sediment sorption will not be a major fate for HMX in surface waters.

The transport of pink water compounds including HMX (4 mg/L) through columns of garden soil (6.5% organic matter content) amended with microbes from activated sludge and anaerobic sludge digest was studied by Greene et al. (1984). Pink water solutions were continuously pumped through the columns, and effluent samples were collected weekly for 110 d. Flow rates were varied and some columns were amended with glucose. Unmetabolized HMX was rapidly recovered in the leachates, indicating little adsorption to soil.

Checkai et al. (1993) collected intact soil-core columns from an uncontaminated area at the Milan AAP to study transport and transformation of munitions chemicals in site-specific soils. The soil was a Lexington silt loam with a 6-in. A horizon containing 16 g/kg organic matter and a B horizon extending from 6 to 27 in. and containing 5 g/kg organic matter. A mixture of munitions simulating open burning/open detonation ash was added to the soil surface. Concentrations were 1000 mg/kg RDX, 1000 mg/kg HMX, 1000 mg/kg 2,4-dinitrotoluene, and 400 mg/kg 2,6-dinitrotoluene. The columns were leached with simulated rainfall over a period of 32.5 wk; controlled tension was applied.

HMX was measurable in leachates throughout the study and averaged 0.4 mg/L. Over time, HMX was found at progressively greater depths within the soil column; by 19.5 wk, it had migrated throughout the full column (detection level, 2.9 mg/kg).

Biotransformation. Biotransformation will be an important pathway for degradation of HMX under organic-rich conditions. No biodegradation of HMX occurred after 15 wk when 4 ppm HMX was added to river waters or river waters with HMX contaminated sediment (Spanggord et al. 1982b). However, the addition of yeast to river water containing 4 ppm of HMX reduced the concentration to <0.1 ppm in 3 d. Spanggord et al. (1982b) found that anaerobic transformation of HMX followed a similar pattern. No degradation occurred in river water, slow degradation occurred in water over HMX-contaminated sediment (from 4 to <0.2 ppm in 91 d), and rapid degradation occurred in the yeast-treated water. Aerobic and anaerobic biotransformation of HMX ultimately produces 1,1-dimethylhydrazine.

In a detailed follow-up study, Spanggord et al. (1983) further investigated the biotransformation of HMX in water. HMX biotransformation occurred under both aerobic and anaerobic conditions in waste stream effluent treated with yeast. Under aerobic conditions a first-order half-life of 17 hr was observed; under anaerobic conditions, the first-order half-life was 1.6 hr. However, conditions in the Holston River environment are not expected to induce noticeable biotransformation of HMX, and thus biotransformation is not expected to contribute to loss of HMX in this environment.

B. Aquatic Toxicology

1. Acute Effects: Invertebrates and Fish. Bentley et al. (1977b) studied the acute effects of HMX on four species of freshwater invertebrates: the midge *Chironomus tentans*, the cladoceran or water flea *Daphnia magna*, the amphipod *Gammarus fasciatus*, and the isopod *Asellus militaris* (Table 51). After 48 hr under static conditions, even the highest tested nominal concentration (32 mg/L) had no adverse effect on any of the organisms. Thus, the 48 hr-EC_{50}, > 32 mg/L, based on immobilization is reported for the invertebrates tested in this study. It should be noted that this concentrations is above the reported solubility limit of HMX (6.6 mg/L at 20 °C).

Bentley et al. (1977b) also studied the effects of HMX on fish tested under static conditions. Test species included bluegill sunfish (*Lepomis macrochirus*), channel catfish (*Ictalurus punctatus*), fathead minnow (*Pimephales promelas*), and rainbow trout (*Oncorhynchus mykiss*). Bluegill sunfish were used to test the effect of changes in temperature, pH, and water hardness. None of these factors had a significant effect. Fathead minnows were used to determine how HMX acute toxicity would vary with life stage. Fathead minnows were most susceptible at 7 d post hatch (96-hr LC_{50} of 15 mg/L). All other life stages for fathead minnows were unaffected by even the highest levels of HMX tested (96-hr LC_{50} values, >32 mg/L). Channel catfish and rainbow trout were similarly unaffected.

Table 51. Acute toxicity of HMX to aquatic invertebrates and fish.[a]

Test species	Test duration	LC_{50} (mg/L) for fish; EC_{50} (mg/L) for invertebrates[b]	Reference
Daphnia magna (water flea)	48-hr	>32	Bentley et al. 1977b
Asellus militaris (sowbug)	48-hr	>32	Bentley et al. 1977b
Gammarus fasciatus (scud)	48-hr	>32	Bentley et al. 1977b
Chironomus tentans (midge)	48-hr	>32	Bentley et al. 1977b
Lepomis macrochirus (bluegill)	96-hr	>32	Bentley et al. 1977b
Ictalurus punctatus (channel catfish)	96-hr	>32	Bentley et al. 1977b
Pimephales promelas (fathead minnow)	96-hr	15	Bentley et al. 1977b
Oncorhynchus mykiss (rainbow trout)	96-hr	>32	Bentley et al. 1977b

[a]Static conditions.
[b]These nominal concentrations exceed the solubility limit of 6.6 mg/L at 20 c°C.

2. *Chronic Effects: Invertebrates and Fish.* One study was located that tested the chronic toxicity of HMX to aquatic organisms. Bentley et al. (1984) tested the effects of long-term HMX exposure on the water flea (*D. magna*) and the fathead minnow (*P. promelas*) under flow-though conditions (6.4 aquarium volume replacements per day for fathead minnows; 4–5 volume replacements per day for water fleas) (Table 52). Fathead minnow embryo-larval stages were exposed to HMX for 32 d. Within 48 hr of fertilization, embryos were suspended in HMX-treated aquariums, and percentage hatch was evaluated at five nominal concentrations of HMX (0.31–5.0 mg/L) and a control. All live larvae were transported to the aquariums upon hatching, and 32 d later the percentage survival, mean total length, and average wet weight were determined. Daphnids were exposed to five nominal concentrations of HMX (0.69–11 mg/L; 80 daphnids per treatment concentration) for a total of 28 d. Weekly survival and offspring production were measured. All toxicity endpoints for both the fathead minnow and water flea were comparable to control levels for all concentrations tested. HMX concentrations were also measured, and were found to be significantly lower than nominal concentrations, especially at the higher nominal HMX levels. For the nominal concentration of 11 mg/L used in the water flea test, measured concentrations ranged from 2.7 to 6.0 mg/L with a mean of 3.9

Table 52. Chronic toxicity of HMX to aquatic invertebrates and fish.[a]

Species	Stage/age	Parameter measured	Parameter response	Concentration (mg/L)	Reference
Daphnia magna (water flea)	Adult	Weekly survival and offspring production; 28 d	No adverse effect	3.9[b]	Bentley et al. 1984
Pimephales promelas (fathead minnow)	Embryo–larvae	Percentage hatch for embryos, percentage survival, mean total length, and average wet weight for larvae; 32 d	No adverse effect	3.3[b]	Bentley et al. 1984

[a]Flow-through tests.
[b]Mean measured concentration.

mg/L, and at the nominal concentration of 5 mg/L used in the fathead minnow test, measured concentrations ranged from 2.3 to 4.9 mg/L with a mean of 3.3 mg/L. Overall, it was concluded that no adverse effects of exposures were seen to *D. magna* at 3.9 mg/L HMX or to *P. promelas* at 3.3 mg/L HMX.

3. Plants. The acute toxicity study of Bentley et al. (1977b) was the only study found that investigated the effects of HMX on aquatic plants. The blue-green algae *Microcystis aeruginosa* and *Anabaena flos-aquae*, the green alga *Selenastrum capricornutum*, and the diatom *Navicula pelliculosa* were tested by measurement of chlorophyll *a* content and culture cell numbers. HMX had no adverse effect on any of the four species of phytoplankton at the tested concentrations; in fact, cell counts or chlorophyll *a* content increased in each case. EC_{50} values were then assumed to be >32 mg/L (Table 53). This concentration is above the level of solubility as indicated by McLellan et al. (1988). Sullivan et al. (1979) summarized and evaluated the HMX aquatic data of Bentley et al. (1977b) and considered the statistical analysis to be questionable. On reanalysis, considering growth in cell counts or chlorophyll *a* content as the endpoint, HMX concentrations as low as 10 mg/L significantly increased growth in three of the algae species and the chlorophyll *a* content in one species.

4. Metabolism and Bioconcentration. No data on HMX metabolism by aquatic organisms were found. Reported log K_{ow} values of 0.26 (Burrows et al. 1989), 0.13 (Tucker et al. 1985), and 0.06 (Jenkins 1989) indicate little potential for partitioning to body lipids; thus, HMX is not expected to bioaccumulate in fish.

C. Aquatic Criteria and Screening Benchmarks

1. Aquatic Organisms. Data were insufficient for calculation of acute and chronic WQC according to USEPA guidelines (Stephan et al. 1985). Guideline data are available for only six of the required eight families, representing planktonic crustaceans (*D. magna*), the family Salmonidae (*O. mykiss*), a second family in the class Osteichthyes (*L. macrochirus* or *I. punctatus*), a third family in the phylum Chordata (*P. promelas*), benthic crustaceans (*G. fasciatus* or *Asellus militaris*), and insects (*C. tentans*). Therefore, Tier II or secondary acute and chronic values were calculated according to USEPA guidance for the Great Lakes System (USEPA 1993a).

It should be noted that the tests conducted by Bentley et al. (1977b) and on which the criteria/benchmarks are based do not follow the methodology outlined by Stephan et al. (1985). The acute experiments performed by Bentley et al. (1977b) were under static conditions, and HMX concentrations were reported as nominal (not measured), being computed based on the dilution of a superstock solution of HMX dissolved in acetone.

Because of the stability and slow removal rates of HMX in the water environment, HMX is not expected to be significantly lost in the time frames used. Thus, the static conditions used by this study should not significantly alter the

Table 53. Toxicity of HMX to algae.[a]

Test species	Test duration	Effect	Concentration (mg/L)[b]	Reference
Microcystis aeruginosa (blue-green alga)	96-hr	No adverse effect	32	Bentley et al. 1977b
Anabaena flos-aquae (blue-green alga)	96-hr	No adverse effect	32	Bentley et al. 1977b
Selenastrum capricornutum (green alga)	96-hr	No adverse effect	32	Bentley et al. 1977b
Navicula pelliculosa (diatom)	96-hr	No adverse effect	32	Bentley et al. 1977b

[a]Static conditions.
[b]These nominal concentrations exceed the solubility limit of 6.6 mg/L at 20 °C.

results. However, nominal concentrations reported in this study were in excess of the solubility limits of HMX (6.6 mg/L at 20 °C). Bentley et al. (1984) reported that measured HMX concentrations averaged only 3.9 mg/L in tests in which the nominal concentration was 11 mg/L. The later study utilized flow-though conditions and used HMX concentrations taken from aquarium measurements in determining HMX levels of concern. For analysis of the acute values taken from Bentley et al. (1977b), concentrations exceeding the known solubility of HMX were used in deriving water quality criteria in accordance with Stephan et al. (1985): "'greater than' values and those which are above the solubility of the test material should be used because rejection of such acute values would unnecessarily lower the FAV by eliminating acute values for resistant species." However, it should be noted that these values based on nominal concentrations may very well be overestimates and thus the confidence in the aquatic criteria/benchmarks is low. Although acute/chronic ratios were available for two families. Daphnidae and Cyprinidae, the "greater than" values for both the acute and chronic data for Daphnidae made the ratio unacceptable. However, the value for Daphnidae, although not definitive, indicates the low toxicity of HMX and will be used to satisfy the data base requirement for a GMAV for Daphnidae.

The SAV is calculated by dividing the lowest GMAV in the database by the SAF. The lowest GMAV was 15 mg/L for *P. promelas*. USEPA (1993a) lists a SAF of 4.0 for use in Tier II calculations when six satisfied data requirements for Tier I calculations are available.

Therefore

$$SAV = \frac{15}{4.0}\ mg/L = 3.75\ mg/L$$

The SMC is one-half of the SAV or 1.875 mg/L.

Values for acute and chronic tests are taken from Tables 51 and 52. Although acute and chronic tests were available for *D. magna* and *P. promelas*, several of the values are NOAELs rather than LOAELs, making the LOAELs "greater than" values. The ACR of > 32 mg/L/ > 3.9 mg/L for *D. magna* is not quantifiable and therefore was not used. Only the chronic value for *P. promelas* is a NOAEL, and the resultant ACR (15 mg/L/ > 3.3 mg/L) value can be considered conservative, i.e., <4.55 (Table 54). Thus, only one of three required experimen-

Table 54. Aquatic acute/chronic ratios (ACR) for HMX.

Species	Acute values (mg/L)	Chronic values (mg/L)	Acute/chronic ratios (ACR)
Fathead minnow	15	>3.3	<4.55
Default ACR			18
Default ACR			18

tally determined ACRs were available according to guidelines for Tier I. Therefore, two assumed ACRs of 18 were added. The SACR is the geometric mean of the three ACRs. Note that since the chronic value endpoints were not seen at the highest concentrations measured, some additional uncertainty in the SACR will result from the use of lower limit toxicity endpoints. The SACR is the geometric mean of 4.545, 18, and 18 or 11.38.

The SCV is the SAV divided by the SACR:

$$SCV = \frac{SAV}{SACR} = \frac{3.75}{11.37} = 0.3296 \text{ mg/L}$$

Aquatic criteria and screening benchmarks are listed in Table 55. Although a SCC was calculated according to Tier II guidelines, it is overly conservative compared to measured concentrations that resulted in no adverse effects in chronic tests with *D. magna* and *P. promelas*. In the absence of data for calculation of a CCC or SCC, the USEPA lists LOEC values. A LOEC is not available from the chronic studies but would be greater than the chronic NOAELs of 3.3 and 3.9 mg/L. Therefore, the LOEC of >3.3 mg/L would be a more realistic interim value until additional toxicity tests are performed.

A FPV was not available. For algal species, however, a test of at least 96-hr duration should be used; concentrations of test material should be measured and the endpoint should be biologically significant (Stephan et al. 1985). In considering protection of plant species, increases in growth will not be considered biologically important, and the lower-limit EC_{50} value of >32 mg/L for all four algae species tested by Bentley et al. (1977b) can be considered an estimate of the FPV.

Table 55. Water quality criteria/screening benchmarks for HMX.

Criterion	Value
Acute water quality criterion	Insufficient data
Chronic water quality criterion	Insufficient data
Secondary acute value	3.8 mg/L
Secondary maximum concentration	1.88 mg/L
Secondary chronic value	0.33 mg/L
Secondary continuous concentration	0.33 mg/L
Lowest chronic value, fish	>3.3 mg/L[a]
Lowest chronic value, daphnids	>3.9 mg/L[a]
Final plant value	>32 mg/L
Sediment quality benchmark (SQB_{oc})	0.47 mg/kg_{oc}[b]

[a]These chronic NOAELS may be more realistic representations of the secondary continuous concentration (SCC) than the calculated value of 0.33 mg/L. These lowest chronic values are based on reproduction (daphnids) and no effects on young sensitive stages of fish and use measured concentrations, whereas the SCC is based on few data and the use of default acute/chronic ratios.

[b]mg chemical/kg organic carbon in sediment.

2. Sediment-Associated Organisms. One study was found that tested the toxicity of HMX to benthic organisms: the midge *C. tentans*, the isopod *A. militaris*, and the amphipod *G. fasciatus* (Bentley et al. 1977b); however, the tests were conducted in 250-ml beakers without sediment and cannot be used to derive a SQB. From these tests, it is known that a sediment pore-water concentration at the limit of HMX solubility would not be acutely toxic to these organisms (see Table 51).

The EqP approach can also be used to calculate a SQB for HMX. Because K_{ow} values are available, a K_{oc} can be calculated and a SQB_{oc} can be determined. Using the mean of the log K_{ow} values cited by Burrows et al. (1989) of 0.16 and the relationship between K_{ow} and K_{oc}, the log K_{oc} is 0.1576 and the K_{oc} is 1.437. Then, the organic carbon-normalized SQB_{oc}, i.e., mg/kg organic carbon (mg/kg_{oc}) is

$$SQB_{oc} = K_{oc} \times SCV$$
$$SQB_{oc} = 1.437 \text{ L/kg} \times 0.3296 \text{ mg/L} = 0.474 \text{ mg/kg}_{oc}$$

If an organic carbon content in the soil of 1% is assumed, the $SQB_{1\%}$ is calculated as

$$SQB_{1\%} = 0.01 \times 1.437 \text{ L/kg} \times 0.3296 \text{ mg/L} = 0.005 \text{ mg/kg}$$

D. Terrestrial Toxicology

1. Mammals. No information on the toxicity of HMX to wildlife was found. Two studies were located that investigated the subchronic toxicity of HMX to laboratory animals for periods greater than 2 wk. The studies are summarized in Table 56.

Everett et al. (1985) incorporated HMX into the diets of male and female Fischer-344 rats for 13 wk. Twenty rats were tested at each dose level for both sexes; males at 0, 50, 150, 450, 1350, and 4000 mg/kg/d and females at 0, 50, 115, 270, 620, and 1500 mg/kg/d. Three deaths were recorded: 1 male in the 150 mg/kg/d group after 9 wk, 1 control group female at wk 13, and 1 female at the highest dose level during the first week. The single death in the group of males receiving 150 mg/kg/d was not considered treatment related as no deaths occurred at the three higher doses. There were no clinical signs of toxicity in the HMX-treated animals, although significant dose-related reductions in mean body weight gain occurred for all animals concurrent with reductions in food consumption early in the study. After wk 5, only animals in the two highest dose groups continued to have depressed weight gains.

Histopathological examinations revealed that males receiving 450 mg/kg/d or more of HMX had a significant incidence of toxic liver changes, with some incidences noted as low as 150 mg/kg/d. In females, the three highest dose levels resulted in significant incidences of tubular kidney changes. Thus, the toxicity of HMX on organs was demonstrated to be sex specific. Hematology, clinical chemistry, urinalysis, and organ weight and pathology examinations

Table 56. HMX toxicity data for mammalian species.

Species	Route	Exposure period	NOAEL (mg/kg/d)	LOAEL (mg/kg/d)	Effect/endpoint	Reference
Rat (males)	Diet	13 wk	50	150[a]	Hepatic effects	Everett et al. 1985
(females)			115	270	Renal effects	
Mouse (males)	Diet	13 wk	30	75	Mortality	Everett and Maddock 1985
(females)			30	250		

[a]No treatment-related mortality.

were conducted on the males and females receiving the highest dose levels, with comparisons made to the control group. Changes occurring in one or both sexes at different time intervals included decreased hemoglobin, hematocrit, and red blood cell count and increases in serum alkaline phosphatase, albumin, and blood urea nitrogen. Organ weights were recorded at all dose levels, and a number of organ weight changes were observed for both males and females at all but the lowest dose levels. However, it is difficult to interpret their toxicological significance because many organ weight changes simply reflected the decreases in body weight.

Everett and Maddock (1985), in a study similar to the 13-wk study, tested the oral toxicity of HMX in $B6C3F_1$ mice. Males were dosed with 0, 5, 12, 30, 75, and 200 mg/kg/d and females with 0, 10, 30, 90, 250, and 750 mg/kg/d, in groups of 20 animals each. Significant mortality resulted: 1 male in the 30 mg/kg/d group, 2 males in the 75 mg/kg/d group, and 13 males in the 200 mg/kg/d group died; 1 control female, 1 female in the 30 mg/kg/d group, 12 females in the 250 mg/kg/d group, and all females in the 750 mg/kg/d group died. Surprisingly, no clinical signs of toxicity were noted in any of the treatment groups. Similarly, no significant changes in body weight or clinical chemistry were observed. The single death in each of the 30 mg/kg/d dose groups was not statistically significant. Histological examination of the highest dosed males and females showed few changes in the liver, kidney, spleen and brain. Lungs were observed to be a dark-red color in some of the highest dosed mice, but lungs were not histologically examined.

2. Birds. No studies were found that tested the toxicity of HMX to birds.

3. Plants. There are no studies that specifically tested the toxicity of HMX to terrestrial plants. However, Simini et al. (1995) evaluated the toxicity of Joliet AAP soil to radishes and cucumbers. The soil samples were contaminated with several munitions, each at a range of concentrations. HMX was found to be very weakly correlated with effects on plant height.

4. Soil Invertebrates. A study was found that tested the toxicity of munitions-contaminated soil (contaminated in part with HMX) and calculated the correlation of HMX concentration with overall soil toxicity. Simini et al. (1995) evaluated the toxicity of Joliet AAP soil to earthworms by determining survival rates and differences between initial and final live weights. HMX was found to be very weakly correlated with these endpoints.

In a preliminary study, Phillips et al. (1993) tested the toxicity of HMX added to artificial soil (1.4% organic matter) to the earthworm *Eisenia foetida* during a 14-d period. Concentrations of HMX were 0, 50, 100, 200, 400, and 500 mg/kg. Although survival was 100% for all tested concentrations, there was an increasing weight loss with increasing concentrations. Weight losses ranged from 6% at the 50 mg/kg concentration to 15%–18% at the three highest concentrations.

5. Soil Heterotrophic Processes. No studies were located that tested the toxicity of HMX to soil microorganisms.

6. Metabolism and Bioaccumulation.

Animals. HMX appears to be poorly absorbed by the gastrointestinal tract (ATSDR 1994). Cameron (1986) exposed B6C3F_1 mice and F344 rats to single oral doses (500 mg/kg) of [^{14}C]HMX to investigate HMX excretion. Four days after exposure to rats, 85% of the radiolabel was excreted in feces (primarily as unchanged HMX), 4% in the urine, and, within 48 hr, 0.5% in exhaled air (as $^{14}CO_2$). Respective values for the mouse were 70%, 3%, and 1%. Only 0.6%–0.7% of the original radioactivity remained in the animals after 4 d. Cameron (1986) found that peak HMX plasma levels only reached <0.1% of the administered dose. A comparison of urine and plasma levels of radioactivity following oral and intravenous administration to rats suggested that <5% of the oral dose was absorbed. Similarly, Everett et al. (1985) found low, non-dose-dependent levels of HMX in plasma of rats following 13 wk of exposure to 50–4000 mg/kg/d and concluded that little of the dose was absorbed. No information was found on the distribution of HMX in the body following an oral dose. Following intravenous administration to the rat, highest concentrations of radioactivity were observed in liver and kidney, and the lowest concentration was in the brain (Cameron 1986).

Few data exist on the metabolism of HMX. Following intravenous administration to the rat, unidentified polar metabolites were present in the tissues, plasma, and urine (Cameron 1986). Everett and Maddock (1985) analyzed the stomach contents of B6C3F_1 mice dosed with 5–750 mg/kg/d for nitrite (which would be formed from cleavage of the ring nitrogen–nitrogen bonds) and did not observe any increases above normal dietary concentrations. It was concluded that HMX was not being metabolized in the stomach. Thus, most HMX ingested will be excreted in the feces as unchanged HMX, with the small amount of absorbed HMX being excreted in urine within a few days (ATSDR 1994).

HMX is rapidly excreted in mammals (ATSDR 1994). Bioaccumulation is not expected to occur. Biomonitoring studies have been conducted at Aberdeen Proving Ground, MD (USACHPPM 1994), and at several other AAPs (see Opresko 1995b for review). HMX was not present in tissues of terrestrial wildlife (deer and small mammals) at or above a detection limit of 0.1 mg/kg.

Plants. No studies were found regarding HMX uptake and metabolism in plants.

E. Terrestrial Criteria and Screening Benchmarks

1. Mammals. Two 13-wk studies of the oral toxicity of HMX to laboratory animals were available for calculation of screening benchmarks for mammalian wildlife. Everett et al. (1985) determined renal and hepatic effects of HMX exposure on male and female F344 rats, and Everett and Maddock (1985) measured HMX-induced mortality in male and female $B6C3F_1$ mice. Because the single deaths in the two groups administered 30 mg/kg/d were not significant, this dose was considered a NOAEL. Considering male and female endpoints from both studies, NOAELs ranged from 30 to 115 mg/kg/d, and LOAELs ranged from 75 to 270 mg/kg/d. Despite using the more severe endpoint of mortality in the mouse study, NOAELs and LOAELs were generally lower than those found based on liver and kidney damage to rats. Mortality was considered the more relevant endpoint for wildlife population level effects; thus, the NOAEL of 30 mg/kg/d (for both males and females) from the mouse study was chosen for calculation of screening benchmarks for mammalian wildlife. Conversely, for humans, an oral RfD of 0.05 mg/kg/d was calculated by dividing the NOAEL of 50 mg/kg/d for male rats (endpoint, liver lesions) by an overall uncertainty factor of 1000 (USEPA 1993e).

The subchronic NOAEL of 30 mg/kg/d in the mouse study was multiplied by 0.1 to derive an estimated chronic NOAEL of 3.0 mg/kg/d. The mean body weight (geometric mean of weekly means) for males in the 30 mg/kg dose group was calculated to be 0.025 kg, and that for females in the 30 mg/kg dose group was 0.023 kg. In calculating NOAEL values for mink, the log K_{ow} of 0.26 was used to estimate the FCM. For this log K_{ow}, the FCM is 1.0. Screening benchmarks for wildlife species are listed in Table 57. Confidence in these benchmarks is low due to the limited number of studies and the variability in HMX toxicity.

2. Birds. No studies were found that tested the toxicity of HMX to birds, so screening benchmarks cannot be derived.

3. Plants. No studies were found that directly tested the toxicity of HMX to terrestrial plants. The correlation of HMX concentration in munitions-contami-

Table 57. HMX screening benchmarks for selected mammalian wildlife species.

Wildlife species	Chronic NOAEL (mg/kg/d)	Screening benchmarks		
		Diet (mg/kg food)	Water (mg/L)	Piscivorous species (mg/L)[a]
Shorttail shrew	3.3	5.6	15	—
White-footed mouse	3.0	20	10	—
Meadow vole	2.6	22	19	—
Cottontail rabbit	1.1	5.6	12	—
Mink	1.2	8.5	12	5.7
Red fox	0.8	8.0	9.6	—
Whitetail deer	0.4	14	6.6	—

[a]Water concentration that incorporates dietary intake from both water and food consumption.

nated soil to toxicity in garden plants calculated by Simini et al. (1995) is not applicable to deriving screening benchmarks.

4. Soil Invertebrates. According to the method of Will and Suter (1995b), a 20% reduction in growth would be a threshold for significant effects. Although concentrations up to 500 mg RDX/kg of artificial soil were not lethal to the earthworm *E. foetida*, weight losses occurred but were <20% at the highest concentration tested, 500 mg/kg (Phillips et al. 1993). In addition, HMX in other soil types may behave differently than in artificial soils, and additional studies are needed before a screening value for earthworms can be determined.

5. Soil Heterotrophic Processes. No studies were found that tested the toxicity of HMX to soil microorganisms, therefore, screening benchmarks cannot be derived.

IX. *N*-Methyl-*N*,2,4,6-Tetranitroaniline

The nitramine, *N*-methyl-*N*,2,4,6-tetranitroaniline or tetryl, has been used since 1906 as a military explosive, primarily as a propellant or detonator charge. Tetryl is no longer manufactured, and since 1973 it has been largely replaced in modern explosive formulations by RDX (Kaye 1980b). Contamination from past activities is localized around AAPs or ordnance works where tetryl was manufactured, used, or stored. Chemical and physical properties are listed in Table 58.

This section on tetryl summarizes the available data on concentrations in environmental media, environmental fate and transport, and ecotoxicity and bioaccumulation data for aquatic and terrestrial species. Information was available on transformation in the environment (photolysis, hydrolysis, and biotransformation) and toxicity to laboratory animal species. A subchronic study with the rat was used to derive screening benchmarks for food and water intake for wildlife species. LOECs for terrestrial plants were also located. No suitable data for derivation of screening benchmarks for aquatic invertebrates or fish, sediment-associated organisms, avian species, soil invertebrates, or soil heterotrophic processes were found.

A. Environmental Fate

1. Sources and Occurrences. Tetryl is a military-unique compound that was manufactured primarily at Joliet AAP. Although it has not been manufactured in the U.S. since 1973, existing stocks are still stored at military installations. At the time of closure, it was estimated that 31,000 lb of tetryl were present in the soil at the Joliet AAP (Small and Rosenblatt 1974). Although not usually detectable in surface waters, concentrations in groundwater ranged up to 67 μg/L; the maximum concentration found in subsurface soil was 84,400 mg/kg. At least 19 additional AAPs have handled tetryl. Wastewaters at production

Table 58. Chemical and physical properties of *N*-methyl-N-2,4,6-tetranitroaniline (tetryl).

Synonyms	N-Methyl-N,2,4,6-Tetranitro Benzenamine N-Methyl-N,2,4,6-tetranitro Aniline 2,4,6-Tetryl 2,4,6-Trinitrophenylmethylnitramine 2,4,6-Trinitrophenyl-*N*-methylnitramine Methyl-2,4,6-trinitro-phenylnitramine Methylpicrylnitramine Nitramine Picrylmethylnitramine Picrylnitromethylamine Pyrenite Tetralite Tetril Trinitrophenylmethylnitramine	ATSDR 1995d; ACS 1998
CAS number	479-45-8	HSDB 1995g
Molecular weight	287.15	Budavari et al. 1996
Physical state	Colorless to yellow crystals	Budavari et al. 1996
Chemical formula	$C_7H_5N_5O_8$	Budavari et al. 1996
Structure	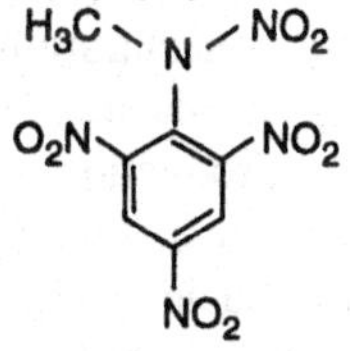	Budavari et al. 1996
Water solubility (20 °C)	75 mg/L	Small and Rosenblatt 1974
(25 °C)	80 mg/L	Yinon 1990
Specific gravity	1.57	Budavari et al. 1996
Melting point	130°–132 °C	Budavari et al. 1996
Boiling point	Explodes at 180°–19 °C	Budavari et al. 1996
Vapor pressure (20 °C)	4×10^{-10} mm Hg	Layton et al. 1987
	1×10^{-8} mm Hg	SRC 1995b
(25 °C)	5.7×10^{-9} mm Hg	Burrows et al. 1989
Partition coefficients		
Log K_{ow}	2 (calculated)	Layton et al. 1987
	1.65 (calculated)	Jenkins 1989
Log K_{oc}	1.69 (calculated)	Burrows et al. 1989
	2.6 (calculated)	SRC 1995b
	3.1–3.5	Hale et al. 1979
Henry's law constant	1.0×10^{-11} atm-m^3/mole	SRC 1995b

sites were released to holding lagoons for settling of solids before being released to rivers and streams. The compound has also been disposed of by ocean dumping. Presently, tetryl may be released to environmental media during Department of Defense demilitarization operations involving exploding or burning. According to Walsh and Jenkins (1992), tetryl is rarely observed in environmental samples because of its instability.

Air. Although tetryl could be released to the air during destruction activities, concentrations in air have not been monitored during these activities (ATSDR 1995d).

Surface Water and Groundwater. According to Small and Rosenblatt (1974), approximately 36 lb of tetryl from each of 12 production lines were discharged daily in wastewater at the Joliet AAP; wastewaters containing 400–460 mg/L were released to drainage ditches. Production ceased at this site in July 1973. Seepage water that was collected in a hole dug close to a drainage ditch at the Joliet AAP in 1973 contained a tetryl concentration of 44 mg/L. However, the concentration in surface water in the ditch was below the limit of detection. A 1988 survey showed a tetryl concentration of 67 μg/L in a sample from a groundwater monitoring well in this area (US Army 1990).

At the Iowa AAP where booster charges were molded from bulk explosives, about 11,000 lb of tetryl were processed (Small and Rosenblatt 1974). Wastewater was transported to a sedimentation pond. At this site, tetryl was present in groundwater below a settling lagoon at a concentration of 46 μg/L (Clear and Collins 1982). However, in another study at this site, tetryl was not detected in surface water or groundwater in the tetryl production area (Deaver et al. 1986).

At the Louisiana AAP, tetryl was found in groundwater below several areas: concentrations were 1.4–53 μg/L (Area P leaching pits), 1.4–5.1 μg/L (burning ground 5), and 0.9–10.2 μg/L (inert landfill area No. 8) (Gregory and Elliott 1987). Concentrations in most samples were below the detection limit of 1.0 μg/L. During a remedial investigation study, tetryl was detected in 2 of 18 wells at this site (Todd et al. 1989), at 1.5 and 3.1 μg/L. Groundwater contamination at the Alabama AAP was confined to the aquifer below the tetryl manufacturing area (ESE 1986). The highest concentration measured was 36.4 μg/L.

Soils. At the Alabama AAP, tetryl was present in surface soil at concentrations of 0.5–6.6 mg/kg (Rosenblatt and Small 1981). In addition, crystalline material suspected of being this contaminant was visible in soils in one area. In another survey at this site, five of seven soil samples collected from the tetryl manufacturing area were contaminated with varying amounts of tetryl, 0.078 to 13,600 mg/kg (ESE 1986). Concentrations in the flashing ground ranged up to 6620 mg/kg (ATSDR 1995d). Subsurface soil close to a drainage ditch at the Joliet AAP contained 84,400 mg/kg soil (Small and Rosenblatt 1974). In 1981, concentrations in two of five soil samples collected from the tetryl production area at Joliet AAP were 23.3 and 38,500 mg/kg (Deaver et al. 1986). At an open burning area at the Picatinny Arsenal, concentrations ranged from 176 mg/kg at the surface to 35 mg/kg at 8–14 in. below the surface (Bauer 1985). Although present in groundwater at the Louisiana AAP, concentrations in soil samples were below the limit of detection of 0.8 ppm (Gregory and Elliott 1987). Concentrations in surface soils from dry lagoons at this site ranged from <0.3 to 42,217 mg/kg (ATSDR 1995d).

Over a 10-yr period, Walsh and Jenkins (1992) collected dried soil samples from 16 Army sites and analyzed the samples for a number of chemicals relating to munitions use and production. Tetryl was found at 2 of these sites; concentrations were 346 mg/kg (one sample) at the Iowa AAP and 0.25–1260 mg/kg (five samples) at the Nebraska Ordnance Works. Tetryl was detected at concentrations of <1000 mg/kg in 4.8% of soil samples taken from open burning grounds at selected military installations (U.S. Army 1986).

2. *Transport and Transformation Processes.* Although tetryl is a stable explosive when stored (Yinon 1990), it undergoes rapid degradation in soil and water.

Abiotic Processes. Tetryl has not been monitored in the ambient air around production and processing sites. Calculated vapor pressures of tetryl of 4×10^{-10} mm Hg at 20 °C (Layton et al. 1987), 5.7×10^{-9} mm Hg at 25 °C (Burrows et al. 1989), and 1×10^{-8} mm Hg (SRC 1995b) indicate a low potential to enter the atmosphere. This low vapor pressure indicates that any tetryl in the ambient atmosphere will exist primarily in the particulate phase and is subject to dry or wet deposition (SRC 1995b). Because it undergoes photolysis in water (Kayser et al. 1984), tetryl in the atmosphere would also be subject to degradation by photolysis. A calculated Henry's law constant for tetryl of 1.0×10^{-11} atm-m^3/mole (SRC 1995b) indicates that volatilization is an insignificant transport pathway from water.

The water solubility of tetryl, 75 mg/L at 20 °C (Small and Rosenblatt 1974), is considered low to moderate and would impede leaching to groundwater. However, picric acid, a primary hydrolysis product of tetryl, with a water solubility of 11,000 mg/L (Layton et al. 1987), may leach to groundwater.

Photolysis and hydrolysis are major environmental transformation processes for tetryl. The rate of degradation via photolysis is at least an order of magnitude faster than hydrolysis (Kayser et al. 1984), making photolysis the dominant degradation process in sunlit waters. Kayser et al. (1984) studied the hydrolysis of tetryl in buffered aqueous solutions at ambient temperature under both light and dark conditions. Under conditions of ambient temperature, a pH of 6, and artificial light, 95.4% of the tetryl (12 mg/L) disappeared from distilled water solutions by 20 d. The authors identified *N*-methylpicramide (41% of the products formed), nitrate (35.2%), nitrite (9.4%), picrate ion (3.9%), and methylnitramine (0.01%) as photolytic and hydrolytic products under these conditions. When tested under similar conditions, but in the absence of light, only 3.2% of the tetryl disappeared in 20 d. Hydrolysis in the dark occurred more rapidly in a borax buffered solution at a pH of 9. Under these conditions, 98.7% of the tetryl disappeared in 90 d. Methylnitramine formation dominated (66%); picrate ion (28%), nitrite (4.1%), nitrate (3.1%), and *N*-methylpicramide (4.1%) were also formed. Methylnitramine appeared to be stable in solutions held in the dark. Additional hydrolysis kinetic studies conducted at several temperatures and a range of pH values were used to estimate the environmental half-life. Based on these studies, the hydrolysis half-life was estimated at 302 d at 20 °C and pH 6.8. In another study, samples of tetryl stored at ambient temperature in the laboratory and protected from light were stable for 33 d at pH ≤4 but were degraded by approximately 50% within 1 d at pH 10.3 (Belkin et al. 1985). In seawater at pH 8, 88% of tetryl hydrolyzed after 101 d, yielding picric acid as a hydrolysis product; the associated half-life was approximately 33 d (Hoffsommer and Rosen 1973).

A solution of tetryl exposed to growth chamber lights progressively decomposed from an initial concentration of 4.73 mg/beaker to 3.64 and 0.83 mg/beaker after 1 and 7 d, respectively. The transformation product *N*-methyl-2,4,6-trinitroaniline was identified in the solution, and the solutions attained a bright yellow color (Harvey et al. 1993). Transformation was slower in the dark, with the concentration decreasing from an initial 4.71 mg/beaker to 4.33 and 4.06 mg/beaker after 1 and 7 d, respectively. *N*-Methyl-2,4,6-trinitroaniline was not found in the beakers held in the dark, and the solutions remained colorless. Fellows et al. (1992) also noted that tetryl solutions that are exposed to light become bright yellow in color.

A number of studies cited the thermal instability of tetryl. During development of analysis methods, heating of tetryl samples at ≥45 °C resulted in decomposition and yielded the thermal degradation products picric acid (the major product), *N*-methylpicramide, picramide, *p*-nitroaniline, *N*-methyl-2-4,6-trinitroaniline, *N*-methyl-*N*,2,4-trinitroaniline, 2,4,6-trinitroanisole, *N*,2,4,6-tetranitroaniline, and 1,3,5-trinitrobenzene (Yasuda 1970; Walsh 1990). A limited amount of thermal degradation may also take place under environmental conditions. *N*-Methyl-*N*-2,4-trinitroaniline is present as a major impurity of production-grade tetryl but is present in only trace amounts after heating (Yasuda 1970).

If released to acidic or neutral soils, tetryl would undergo slow hydrolysis; hydrolysis would be more rapid in alkaline soils. Although K_{oc} values of 49 (Burrows et al. 1989) and 406 (SRC 1995b) have been calculated, the latter based on a water solubility of 75 mg/L (Small and Rosenblatt 1974), higher values have been estimated based on measured soil–water partition coefficients. Using a shake-flask procedure, Hale et al. (1979) measured the soil–water partition coefficient (K_p) for tetryl using four soil types ranging in organic carbon content from 0.39% to 2.2%. The K_p values ranged from 7.6 to 35.3 and K_{oc}, calculated by dividing the K_p values by the respective average organic carbon contents, ranged from 1357 to 2948. These K_{oc} values were higher than predicted from solubility and melting point, indicating to the authors that the methylnitroamino group may bind to soil components. All these K_{oc} values indicate that low to moderate leaching will occur in soil. However, hydrolysis in moist soil will influence the amount of tetryl or tetryl products reaching groundwater. The hydrolysis product, picric acid, may also form complexes with metal ions in the soil (Layton et al. 1987).

Hale et al. (1979) studied the transport of tetryl through soil columns, and Kayser and Burlinson (1988) analyzed soil and water samples from these experiments. ^{14}C-Labeled tetryl was added to the surface layer of 5 × 61 cm soil columns contained in steel pipes; four different types of soil were tested. The columns were irrigated with water (5 cm/wk), and leachate was collected over a 6-mon period. No tetryl was recovered in the leachate. Analysis of the leachate by high-performance liquid (HPLC) and thin-layer chromatography revealed the presence of trace amounts of picric acid and nonvolatile, highly polar products.

Three types of soils with organic matter content ranging from 0.5% to 7.2% were amended with 60 mg/kg tetryl uniformly radiolabeled with ^{14}C and incubated for 60 d (Fellows et al. 1992; Harvey et al. 1992). Both nonsterile and sterile soils, moistened to 67% field capacity, were used. Tetryl was rapidly transformed in both sterile and nonsterile soils, with only 43%–62% of unaltered tetryl recoverable from the soil immediately after amendment. In nonsterile soils, <8% of the parent tetryl was recoverable after 11 d, and by 30 d unaltered tetryl was no longer present. The primary transformation product was *N*-methyl-2,4,6-trinitroaniline; a minor transformation pathway involved direct ring nitro reduction with formation of an aminodinitrophenylmethylnitramine isomer. By 60 d, 21%–36% of the radiolabel was extractable from the soil and 43%–58% of the label remained in the soil. The amount of radiolabel that was initially bound to the soils (d 0) was related to the amount of organic carbon in the soils and cation-exchange capacity (CEC), with highest binding to a clay loam (7.2% organic carbon; 38.2 meq/100 g CEC). In soils sterilized with gamma radiation, transformation was much slower than in the nonsterile soils and the percent unextractable radiolabel in the soil was much lower. The authors acknowledged that the gamma-radiated soils might not have been sterile.

As part of the same study (Fellows et al. 1992; Harvey et al. 1992), volatile organic compounds and $^{14}CO_2$ evolution from the three soils were measured. No volatile organic compounds were evolved from any soil type, sterile or nonsterile; $^{14}CO_2$ evolution did occur from both soils with greater amounts released from the nonsterile soils. Although the fractional amount of the total label evolved as CO_2 was not measured, it was estimated to be ~9% over the 60-d period. Tetryl, *N*-methyl-2,4,6-trinitroaniline (the primary transformation product), a dinitroaminophenylethylnitramine isomer, and two unidentified metabolites were found in the soils.

Biotransformation. No information on microbial degradation was located, but microbial reduction of nitro groups to amines readily takes place (Layton et al. 1987; Fellows et al. 1992). In the soil studies conducted by Fellows et al. (1992) and discussed in Section II.B.1, it is not clear to what extent the reduction and mineralization of tetryl resulted from microbial activity or was an inorganic process. Furthermore, the primary transformation product, *N*-methyl-2,4,6-trinitroaniline, may have been an artifact of the extraction and analysis procedure.

B. Aquatic Toxicology

1. Toxicity. No data concerning the toxicity of tetryl to aquatic organisms were located. In natural, sunlit waters, tetryl would be transformed via photolysis and hydrolysis to several degradation products including picric acid. A survey of the literature on the toxicity of picric acid to aquatic organisms by Burrows and Dacre (1975) found that the concentration below which picric acid was nontoxic for fish (minnow, *Phoxinus phoxinus*) is 30 mg/L (Grindley 1946). Phytotoxicity occurred at higher concentrations. The 96-hr "toxicity threshold" of picric acid for the green alga *Scenedesmus quadricauda* was 240 mg/L (Bringmann and Kuhn 1959). A concentration of 1750 mg/L halved the growth rate of the aquatic vascular plant *Lemna minor* (Simon and Blackman 1953).

2. Metabolism and Bioconcentration. No studies on metabolism or bioconcentration in aquatic organisms were located. The K_{ow} of a chemical (concentration in octanol/concentration in water) can be used to estimate a bioconcentration factor (BCF). Although a K_{ow} for tetryl has not been measured in experimental systems, it can be estimated from solubility data, chemical structure, and reverse-phase HPLC retention time. A mean log K_{ow} for tetryl of 2 (1.5 from reverse-phase HPLC and 2.4 from solubility) was estimated by Layton et al. (1987). The authors used the log K_{ow} value to calculate a bioconcentration factor for fish [(mg chemical in fish/kg body weight of fish)/(mg chemical in water)/(kg water)] of 15. Also using reverse-phase HPLC retention time, Jenkins (1989) calculated a log K_{ow} of 1.65. Based on a water solubility of 75 mg/L and the regression equation of Lyman et al. (1982), SRC (1995b) calculated a BCF of 54. These values suggest a low potential for significant bioconcentration in aquatic organisms.

C. Aquatic Criteria and Screening Benchmarks

No acute or chronic studies on the toxicity of tetryl to invertebrates or fish were located. Therefore, water quality criteria and screening benchmarks could not be derived. No data on the toxicity of spiked sediments or the interstitial water from spiked sediments to sediment-associated organisms were located for this chemical. Although tetryl does not significantly ionize in water (Layton et al. 1987), it strongly binds to soils (Hale et al. 1979) and plant and animal tissues (Lakings and Gan 1981), and thus the EqP approach for calculation of a sediment quality criterion cannot be applied.

D. Terrestrial Toxicology

1. Mammals. No subchronic or chronic studies on the toxicity of tetryl using mammalian wildlife were located. Human exposure has resulted in contact dermatitis, discoloration of the skin, respiratory irritation, and anemia. Symptoms of headache, fatigue, irritability, and insomnia have been reported in workers engaged in tetryl production or processing (Yinon 1990). Studies with laboratory animals have identified the liver and kidneys as target organs of systemic tetryl toxicity.

In an early study, Daniele (1964) and Fati and Daniele (1965) studied the effects on rabbits exposed to tetryl by gavage for 6–9 mon. No gross or histological alterations were observed in the lungs, hearts, or gastrointestinal mucosa of 12 rabbits thus treated with 125 mg/kg/d (Fati and Daniele 1965). The rabbits did show signs of a "coagulation disorder" starting 120 d after the initial exposure (Daniele 1964), accumulation of hematic pigments in their spleen (Fati and Daniele 1965), and swollen hepatocytes (Fati and Daniele 1965). Four rabbits treated for 9 mon also experienced liver damage (hepatocyte necrosis and hyperplasia of the Kuppfer cells) and slight congestion of the kidneys with cloudy swelling and vacuolar degeneration of the convoluted tubules (Fati and Daniele 1965). The 6 rabbits treated for 6 mon did not exhibit kidney congestion. A control group was not used in this study.

Parmeggiani et al. (1956) administered a dose of 50 mg/kg/d to rats for 3 mon. Histological examination revealed slight degenerative changes in the liver and kidneys.

Two recent studies, one a short-term screening study and the other a subchronic study, on the toxicity of tetryl utilizing laboratory rats were located. In the 14-d screening study, Reddy et al. (1994b) administered 0, 500, 2000, 2500, or 5000 mg tetryl/kg diet to 5 male and 5 female Fischer-344 rats. Changes (relative to the control group) occurred in body weights, relative organ weights, and some hematology and clinical chemistry parameters in the higher dose groups. The NOAEL was 500 mg tetryl/kg diet. In the subchronic study, groups of 10 male and 10 female Fischer-344 rats were administered 0, 200, 1000, or 3000 mg tetryl/kg diet for 90 d (Reddy et al. 1994c). Calculated doses were 0, 13, 62, and 180 mg/kg/d for males and 0, 14, 69, and 199 mg/kg/d for females. Food consumption was reduced in all groups throughout the study and resulted in significantly reduced body weights in the 3000 mg tetryl/kg diet (males and females) and 1000 mg tetryl/kg diet groups (females). Relative organ weights were significantly increased in both sexes: kidney (1000 and 3000 mg tetryl/kg diet groups), liver (1000 and 3000 mg tetryl/kg diet groups), and spleen (3000 mg/kg tetryl dose groups). A significant decrease in red blood cell count and a significant increase in reticulocyte count occurred in the 3000-mg dose groups of both sexes at 90 d. Methemoglobin levels were significantly increased in both sexes receiving 1000 and 3000 mg tetryl/kg diet at 45 and 90 d, and hemoglobin was significantly decreased in both sexes in these dose groups at 90 d. Histological changes were observed in the spleen (pigment deposition and erythroid cell hyperplasia in both sexes receiving 3000 mg tetryl/kg diet) and kidney (tubular degeneration and cytoplasmic droplets in males receiving 1000 or 3000 mg tetryl/kg diet). All these effects were absent or not significant in the groups receiving 200 mg/kg/d.

2. Birds. No studies were found that tested the toxicity of tetryl to birds.

3. Plants. Screening studies addressed the toxicity of tetryl to several plant species including *Phaseolus vulgaris* (bush bean), *Triticum aestivum* (wheat), and *Bromus mollis* (blando brome) grown in solutions containing up to 10 mg/L tetryl (near the limit of tetryl solubility under the conditions of the study) or in soil containing up to 75 mg/kg tetryl (Fellows et al. 1992). The 10 mg/L solution concentration was based on a previous, unreported study in which the authors found no or minimal phytotoxic effects at that concentration. In the later screening study, no phytotoxicity was reported in plants cultured in the 10 mg/L solution. However, the study lasted only 10 d and by d 4 the concentration of tetryl in solution was near or below the detection limit of 0.1 mg/L.

Plants were grown for 70 d in soil containing concentrations of 0, 10, 25, 50, or 75 mg/kg (Fellows et al. 1992). There was a slight reduction in growth for bush beans and blando brome at all concentrations, but differences were not significant. The authors considered concentrations up to 50 mg/kg nontoxic for wheat and blando brome grown in any of the three soil types. In the same study, plants grown at a single concentration of 25 mg/kg for 60 d showed variable results, with statistically significant reductions in fresh weights for some plants in some soils. At a level of significance of $p \leq 0.01$, only total fresh weight for wheat grown in Burbank soil at a concentration of 25 mg tetryl/kg soil was decreased compared to the control: 7.43 g versus 10.3 g.

4. Soil Invertebrates. No studies were found that tested the toxicity of tetryl to soil invertebrates.

5. Soil Heterotrophic Processes. No specific studies on the toxicity of tetryl to soil microorganisms were located. In soil degradation studies, a concentration of 60 mg/kg did not appear to impede degradation of tetryl; however, there was no indication that the degradation process was a biological one (Fellows et al. 1992). Tetryl was very toxic to the marine luminescent bacterium *Vibrio fischeri* with a 30-min EC_{50} (50% reduction in luminescence) of 0.45 mg/L (Drzyzga et al. 1995).

6. Metabolism and Bioaccumulation.

Animals. No data on absorption following inhalation or topical administration or on metabolism following oral, inhalation, or topical administration were located. Systemic effects following oral administration indicate that tetryl (or its metabolites) is absorbed and distributed to the tissues and organs (Fati and Daniele 1965; Reddy et al. 1994b,c). Picramic acid, a metabolite of tetryl, was found in the urine of seven rabbits orally administered 100 mg tetryl/d (32.3–40.0 mg tetryl/kg/d) for as long as 30 d (Zambrano and Mandovano 1956). Picramic acid was not detected in the urine of two control rabbits. The data support the hypothesis that tetryl is metabolized to picric acid by removal of the methylnitramine moiety and further metabolized to picramic acid by reduction of a nitro group.

No data on bioaccumulation of tetryl in animals were located. Because tetryl is metabolized and excreted in mammalian species, bioaccumulation is unlikely to occur.

Plants. Transformation of tetryl is dramatic in the presence of plant roots, indicating a surface catalytic process (Cataldo et al. 1993). In the presence of plant roots, tetryl in solution was rapidly transformed to *N*-methyl-2,4,6-trinitroaniline and a dinitroaminophenylmethylnitramine isomer.

The uptake and chemical fate of uniformly ring-labeled ^{14}C-tetryl in plants was evaluated using bush bean, wheat, and blando brome (Fellows et al. 1992; Cataldo et al. 1993; Harvey et al. 1993). Plants were grown in hydroponic solutions at concentrations of 1, 2.5, 5, and 10 mg/L. Tetryl was rapidly taken up by all three plants; by 7 d, >90% of the absorbed radioactivity was present in the roots and <10% was present in the shoots or leaves (in the case of bush beans). Of numerous metabolites extracted from the plant tissues, *N*-methyl-2,4,6-trinitroaniline and *N*-methyl-dinitroamimoaniline were tentatively identified. Much of the radioactivity remained in a nonextractable form.

Plant uptake studies are complicated by the fact that transformation takes place at the plant–soil surface (Cataldo et al. 1993). Comparisons were made of chemical transformations occurring in hydroponic culture solutions maintained over time with and without light, aeration, and plant roots. While transformation occurred in the presence of light and/or aeration, it was more rapid and more extensive when plant roots were present. The presence of *N*-methyl-2,4,6-trinitroaniline and possibly dinitroaminophenylethylnitramine in the solution and the depletion of the radiolabel in the solution after all tetryl had been transformed indicated that plant absorption of both parent compound and transformation products may take place.

When bush bean plants were grown in nutrient solutions containing ^{14}C-labeled tetryl and shielded from light, approximately 42% of the available radioactivity was accumulated within the plant at 7 d (Fellows et al. 1992; Harvey et al. 1993). Radioactivity continued to be accumulated by the plant even after tetryl was no longer detectable in the solutions, indicating that tetryl transformation products as well as tetryl were absorbed by the plant roots.

Uptake of tetryl by plants grown in soil is dependent on soil type (sand > silt ≫ organic soil) and plant species (Fellows et al. 1992). Uptake of tetryl by plants grown for 60 d in soil containing 25 mg tetryl/kg soil ranged from 90 mg tetryl equivalents (tetryl plus metabolites) per plant (fresh weight) for wheat grown in soil with 7.2% organic carbon content to 406 mg tetryl equivalents for blando brome grown in soil with 0.5% organic carbon content. Tetryl residues primarily accumulate in plant roots; approximately 80% of tetryl taken up by mature bush beans was found in plant roots, approximately 17% was in leaves and stems, and approximately 3% was in seeds and pod. In wheat and blando brome, 16%–20% of the accumulated radiolabel was found in the shoot tissues.

E. Terrestrial Criteria and Screening Benchmarks

1. Mammals. Based on hepatic and renal effects in male rabbits (Fati and Daniele 1965), a chronic RfD for humans of 0.01 mg/kg/d was derived (USEPA 1995b). A more recent subchronic study was available for calculating a safe exposure level or NOAEL for wildlife exposed to tetryl in environmental media. In this study (Reddy et al. 1994c), a subchronic oral NOAEL of 13 mg/kg/d was identified in Fischer-344 laboratory rats; the ecologically relevant endpoints of survival and reproduction (no testicular effects were observed) were not adversely affected. The subchronic NOAEL was multiplied by an uncertainty factor of 0.1 to derive the chronic NOAEL. The chronic NOAEL of 1.3 mg/kg/d was used to calculate NOAEL values (by normalization to the body weight of each wildlife species) for food and water intake for the shorttail shrew, white-footed mouse, meadow vole, cottontail rabbit, mink, red fox, and whitetail deer (Table 59). The mean body weight (geometric mean of weekly means for the 13-wk test period) for males in the 13 mg/kg dose group was calculated to be 258 g. The BCF was estimated from the mean log K_{ow} of 1.85 by the equation of Lyman et al. (1982).

2. Birds. No studies were found that tested the toxicity of tetryl to birds.

3. Plants. In the absence of a methodology for development of criteria for terrestrial plants, LOECs from the literature were used to screen chemicals of potential concern for phytotoxicity. Three types of plants (wheat, bush bean, and blando brome) were able to grow for 7 d in hydroponic solutions containing 10 mg/L tetryl (Fellows et al. 1992). This exposure time was too short to be meaningful, and confidence in this benchmark is therefore low. In 70-d growth studies using three species of plants, the LOEC for growth of wheat was 25 mg tetryl/kg soil (Fellows et al. 1992). Although based on one study, three types of

Table 59. Tetryl screening benchmarks for selected mammalian wildlife species.

Wildlife species	Chronic NOAEL (mg/kg/d)	Screening benchmarks		
		Diet (mg/kg food)	Water (mg/L)	Piscivorous species (mg/L)[a]
Shorttail shrew	2.6	4.4	12	—
White-footed mouse	2.4	16	8.0	—
Meadow vole	2.0	18	15	—
Cottontail rabbit	0.9	4.5	9.1	—
Mink	0.9	6.8	9.4	0.4
Red fox	0.6	6.4	7.6	—
Whitetail deer	0.3	11	5.2	—

[a]Water concentration that incorporates dietary intake from both water and food consumption.

plants and three types of soils were used, making confidence in the screening benchmark "moderate." The benchmarks are listed in Table 60.

4. Soil Invertebrates. No studies were found that tested the toxicity of tetryl to soil invertebrates; therefore, screening benchmarks cannot be derived.

5. Soil Heterotrophic Processes. No studies were found that tested the toxicity of tetryl to soil microorganisms; therefore, screening benchmarks cannot be derived.

X. Discussion and Conclusions

The aquatic and terrestrial criteria/screening benchmarks calculated for eight nitroaromatic munition compounds are summarized in Table 61. The quantity and quality of the data used to derive these values varied considerably among the chemicals. Because of the need to estimate the potential environmental toxicity of these chemicals, screening benchmarks were sometimes calculated with less than ideal data. For all chemicals except tetryl, data were sufficient to derive aquatic screening criteria or benchmarks. For some of the chemicals, a sparse database for aquatic toxicity led to the use of default values in the calculations and resulted in conservative benchmarks. When one or more subchronic or chronic studies with several species of test animals were available, derived values were less conservative and there was greater confidence in the values. With the exceptions of DNA and 2-ADNT, data were sufficient to derive chronic oral NOAELs for terrestrial mammals.

USEPA methods and guidelines were used to derive screening values for aquatic organisms, both open water and sediment-associated, and terrestrial mammals. In the absence of guidelines, LOECs were identified for terrestrial plants, soil organisms, and soil heterotrophic processes. As noted, confidence in the screening benchmarks varies with the quantity and quality of the data available for each chemical. Regardless of the confidence in the derived values, benchmarks should be considered on a site-specific basis.

Chemical and physical properties and environmental fate data were available

Table 60. Tetryl screening benchmarks for terrestrial plants and invertebrates.

Screening benchmark	Value
Plants, solution	10 mg/L[a]
Plants, soil	25 mg/kg[a]
Soil invertebrates	Insufficient data
Soil microbial processes	Insufficient data

[a]Based on a single study.

Table 61. Summary of aquatic and terrestrial criteria/screening benchmarks for nitroaromatic munition chemicals.

	Water and sediment quality criteria screening benchmarks			Terrestrial screening benchmarks		
Chemical	Acute[a] (mg/L)	Chronic[a] (mg/L)	Sediment[b] (mg/kg)	Test species chronic oral NOAELs (mg/kg/d)	Plant LOECs (mg/L solution) or (mg/kg, soil)	Invertebrate LOECs (mg/kg soil)
2,4,6-Trinitrotoluene (TNT)	0.57	0.09	9.2	1.6	5 mg/L 30 mg/kg	140[c] 200[d]
1,3,5-Trinitrobenzene (TNB)	0.03	0.01	0.24	6.7[e]	ND	ND
1,3-Dinitrobenzene (DNB)	0.11	0.02	0.67	0.11	ND	ND
3,5-Dinitroaniline (DNA)	0.23	0.06	ND	ND	ND	ND
2-Amino-4,6-dinitrotoluene (2-ADNT)	0.18	0.02	ND	ND	80 mg/kg[f]	ND
RDX	0.70	0.19	1.3	7.0	100 mg/kg	ND
HMX	1.88	0.33	0.47	3.0	ND	ND
Tetryl	ND	ND	ND	1.3	10 mg/L 25 mg/kg	ND

ND, no or insufficient data.
[a]Calculated according to USEPA Tier I (2,4,6-trinitrotoluene) or Tier II guidelines (other chemicals).
[b]mg chemical/kg organic carbon in the sediment; calculated according to USEPA guidelines.
[c]Value for earthworm.
[d]Value for soil invertebrates.
[e]Based on a subchronic study with the white-footed mouse.
[f]Value is a no-observed-effect concentration.

for all the compounds. In most cases, fate predicted on the basis of chemical and physical properties supported laboratory and field studies. However, because laboratory studies are usually conducted under ideal conditions, site-specific conditions such as belowground burial, exposure to weathering, and soil properties need to be considered when predicting environmental fate. Although most of these munitions compounds can be transformed or biodegraded in the environment and some of them sorb to soil, their presence in groundwater years after manufacture indicates some environmental persistence. The following have been found in groundwater at military sites: TNT, TNB, DNB, 2-ADNT, RDX, HMX, and tetryl.

The aquatic toxicity of TNT has been extensively studied, and sufficient data were available to derive WQC according to USEPA Tier I guidelines. Tests were conducted with invertebrates and fish for both acute and chronic time periods for TNB, DNB, DNA, RDX, and HMX. Results of acute tests with an invertebrate and fish species were available for 2-ADNT, but no chronic tests were located. Therefore, USEPA Tier II guidelines were used to derive screening benchmarks for all chemicals except TNT. No aquatic toxicity data were located for tetryl, a compound that rapidly undergoes photolysis and hydrolysis. However, because tetryl has been found in soil and groundwater at production and processing sites, aquatic toxicity tests are needed to assess potential ecological effects.

Screening benchmarks for sediment-associated organisms were calculated only for the nonionic organic compounds TNT, TNB, DNB, RDX, and HMX. No data were located for tetryl. Additional methods are available for other classes of chemicals including metals and ionic compounds, but these are not discussed in this report.

For terrestrial mammals, chronic oral feeding studies with laboratory rats or mice were available for TNT, TNB, and RDX. However, a subchronic study was used to derive the oral NOAEL for TNT because the endpoint of testicular atrophy in the subchronic study was more relevant for population-level effects than the endpoints cited in the two chronic studies. The chronic study with TNB showed that effects seen in a subchronic study were reversible and not life shortening. Oral administration of TNB in two subchronic studies, one with the laboratory rat and the other with the white-footed mouse, demonstrated that a wildlife species, in this case the white-footed mouse, can be more resistant to the toxic effects of chemicals than inbred laboratory strains of mammals. Three chronic oral studies and a teratogenicity/reproductive study were located for RDX, making confidence in the screening benchmark high. One or more subchronic oral studies were available for DNB, HMX, and tetryl, and no oral studies were available for DNA and 2-ADNT. Although chronic oral NOAELs were calculated in this report, less conservative values (LOAELs) can be calculated with the same data by choosing different endpoints or by not applying the LOAEL to NOAEL uncertainty factor. The present screening values are less conservative than similar reference values derived for humans, because effects that do not threaten population abundance can be tolerated in the environment and no animal to human uncertainty factor need be applied.

TNT is present in soil and sediment at former production and handling facilities at concentrations up to 87,000 and 711,000 mg/kg, respectively. It is also present in surface water and groundwater. The primary environmental fate mechanism is photolysis; biodegradation also takes place in the environment. Based on an extensive aquatic database, including tests with eight different families, a chronic WQC of 0.09 mg/L was calculated. Using the EqP method, a SQB of 9.2 mg TNT/kg organic carbon in the sediment was calculated. Using a reproductive endpoint in a subchronic study with the laboratory rat, a chronic oral NOAEL of 1.6 mg/kg/d was calculated.

TNB, an impurity and environmental degradation product of TNT, has been found in soil and groundwater at former TNT production sites at concentrations up to 67,000 mg/kg and 7.7 mg/L, respectively. Data on transport and transformation processes indicate that this chemical is resistant to photolysis and hydrolysis. Based on a moderate database including information on one invertebrate and four species of fish, secondary acute and chronic aquatic screening benchmarks of 0.03 and 0.01 mg/L were calculated. Using the EqP method, a SQB of 0.24 mg of TNB/kg organic carbon in the sediment was calculated. Two subchronic studies, one utilizing a wildlife species, and one chronic study were available for calculation of screening benchmarks for oral intake for mammalian species. Based on the subchronic study with the white-footed mouse, the highest chronic oral NOAEL, 6.7 mg/kg/d, was identified as a screening benchmark. This value is supported by reproductive and developmental toxicity studies in which oral administration to laboratory animals resulted in generally negative findings.

DNB is formed during the production of TNB and from the photolysis of 2,4-dinitrotoluene, another by-product of TNT manufacture. Few data on transport and transformation processes in the environment were available. This chemical appears to be resistant to photolysis and hydrolysis, but is subject to microbial degradation. Based on a moderate database including tests with one invertebrate and four species of fish, secondary acute and chronic aquatic screening benchmarks of 0.11 and 0.02 mg/L, respectively, were calculated. Using the EqP method, a SQB of 0.67 mg of DNB/kg organic carbon in the sediment was calculated. Three subchronic studies utilizing the laboratory rat were evaluated for calculation of a chronic oral NOAEL. The toxic endpoint in all three studies was the same, testicular degeneration/sperm production. The calculated chronic oral NOAEL is 0.11 mg/kg/d.

Few data on DNA were located. This TNT production by-product is formed either through the nitration of toluene or through the chemical or microbial degradation of nitrated toluenes. It is also formed in the environment by the bacterial reduction of TNB. No data on concentrations in the environment were located. Based on a moderate database including acute tests with one invertebrate and four species of fish and chronic tests with one invertebrate and one species of fish, secondary acute and chronic aquatic screening benchmarks of 0.23 and 0.06 mg/L, respectively, were calculated. Because of the potential ionization of the amine group of DNA, a SQB based on the EqP method was not derived. No data on the toxicity of DNA to mammalian species were located.

The TNT degradation product 2-ADNT is moderately persistent in the environment. Acute aquatic toxicity tests were available for one species of invertebrate and one species of fish. Based on this sparse database, the acute and chronic aquatic screening benchmarks are 0.18 and 0.02 mg/L, respectively. Because of the potential ionization of the amine group of 2-ADNT, a SQB based on the EqP method was not derived. No mammalian toxicity data were located.

The explosive RDX can be considered persistent in the environment as it is not amenable to hydrolysis or microbial degradation. The primary degradation pathway is via photolysis. Concentrations in soil at military sites range up to 74,000 mg/kg, and RDX has been found in groundwater below contaminated sites. Although most of the aquatic data on RDX were the result of studies by a single investigator, the data were extensive as four species of invertebrates and four species of fish were used in acute studies and two species of invertebrates and two species of fish were used in chronic studies, one of which involved two generations. Acute LC_{50} values for fish derived by two different investigators were in close agreement. Toxicity data on several species of algae indicate that RDX is toxic only as it approaches its limit of solubility. This extensive database results in high confidence in the aquatic screening benchmarks derived using USEPA Tier II guidelines. The chronic screening benchmark is 0.19 mg/L, and the SQB is 1.3 mg RDX/mg organic carbon in the sediment.

Data were available on the metabolism and toxicity of RDX to mammalian species. Following ingestion by mammals, RDX is extensively absorbed, distributed throughout the body, and metabolized in the liver. Metabolism is extensive as indicated by excretion of 80% of a single ^{14}C-radiolabeled dose in the rat within 4 d. Three chronic studies and one teratogenicity/reproductive toxicity study utilizing laboratory animals were available for calculation of screening benchmarks for mammalian species. From a two-generation oral study with the mouse, a NOAEL of 7.0 mg/kg/d, based on testicular degeneration, was identified.

The explosive HMX is persistent in the environment. The primary degradation pathway is photolysis. Concentrations of 4.2 mg/L in groundwater and 5700 mg/kg in soil have been measured. All aquatic toxicity tests were performed by the same investigator. These studies, utilizing four species of invertebrates and four species of fish in acute tests and one species of invertebrate and one species of fish in chronic tests, determined that HMX is not toxic at its limit of solubility. Acute and chronic aquatic screening benchmarks of 1.88 and 0.33 mg/L were derived. A SQB of 0.47 mg of HMX/kg organic carbon in the sediment was calculated. In mammalian studies, HMX is poorly absorbed following oral ingestion and a chronic NOAEL of 3.0 mg/kg/d was calculated; the population-related endpoint was mortality.

Environmental degradation mechanisms for the military-unique compound tetryl are both photolysis and hydrolysis and tetryl may be considered nonpersistent. Concentrations to 0.2 mg/L in groundwater and 84,400 mg/kg in soil have been measured. No data on aquatic toxicity were located. A single subchronic study with the laboratory rat was available to calculate a terrestrial mammalian oral screening benchmark. In this study, a chronic oral NOAEL of 1.3 mg/kg/d was identified.

Most of the screening values proposed in this report have been reviewed by individuals at ORNL who are knowledgeable about the toxicity of these compounds and the processes of deriving screening benchmarks. However, no federal or state agency or group has approved these proposed values. These screening benchmarks are based on the best data available at this time. These present values are not static, but can be updated as additional studies are conducted and more data become available. For example, when this review was first undertaken, two subchronic studies involving TNB toxicity to mammalian species were available. In 1996, a chronic study was completed that showed that the testicular toxicity observed in the rat in one of the subchronic studies was reversed. As a result of this study, a higher NOAEL was indicated, and the application of a subchronic to chronic uncertainty factor became unnecessary.

Summary

Available data on the occurrence, transport, transformation, and toxicity of eight nitroaromatic munition compounds and their degradation products, TNT, TNB, DNB, DNA, 2-ADNT, RDX, HMX, and tetryl were used to identify potential fate in the environment and to calculate screening benchmarks or safe environmental levels for aquatic and terrestrial organisms. Results of monitoring studies revealed that some of these compounds persist at sites where they were produced or processed. Most of the compounds are present in soil, sediment, and surface water or groundwater at military sites. Soil adsorption coefficients indicate that these chemicals are only moderately adsorbed to soil and may leach to groundwater.

Most of these compounds are transformed by abiotic or biotic mechanisms in environmental media. Primary transformation mechanisms involve photolysis (TNT, RDX, HMX, tetryl), hydrolysis (tetryl), and microbial degradation (TNT, TNB, DNB, DNA, 2-ADNT, and HMX). Microbial degradation for both nitro and nitramine aromatic compounds involves rapid reduction of nitro groups to amino groups, but further metabolism is slow. With the exception of DNB, complete mineralization did not usually occur under the conditions of the studies. RDX was resistant to microbial degradation.

Available ecotoxicological data on acute and chronic studies with freshwater fish and invertebrates were summarized, and water quality criteria or ecotoxicological screening benchmarks were developed. Depending on the available data, criteria/benchmarks were calculated according to USEPA Tier I or Tier II guidelines. The munitions chemicals are moderately to highly toxic to freshwater organisms, with chronic screening values <1 mg/L. For some chemicals, these low values are caused by inherent toxicity; in other cases, they result from the conservative methods used in the absence of data. For nonionic organic munitions chemicals, sediment quality benchmarks were calculated (based on K_{ow} values and the final chronic value) according to USEPA guidelines. Available data indicate that none of the compounds is expected to bioconcentrate.

In the same manner in which reference doses for humans are based on studies with laboratory animals, reference doses or screening benchmarks for wildlife may also be calculated by extrapolation among mammalian species. Chronic NOAELs for the compounds of interest were determined from available laboratory studies. Endpoints selected for wildlife species were those that diminish population growth or survival. Equivalent NOAELs for wildlife were calculated by scaling the test data on the basis of differences in body weight. Data on food and water intake for seven selected wildlife species—short-tailed shrew, white-footed mouse, meadow vole, cottontail rabbit, mink, red fox, and whitetail deer—were used to calculate NOAELs for oral intake. In the case of TNB, a comparison of toxicity data from studies conducted with both the white-footed mouse and the laboratory rat indicates that the white-footed mouse may be more resistant to the toxic effects of chemicals than the laboratory rat and may further indicate the lesser sensitivity of wildlife species to chemical insult. Chronic NOAEL values for the test species based on the laboratory studies indicate that, by the oral route of exposure, TNB and RDX are not highly toxic to mammalian species. However, as seen with TNB, values are less conservative when chronic studies are available or when studies were conducted with wildlife species. Insufficient data were located to calculate NOAELs for avian species.

In the absence of criteria or guidelines for terrestrial plants, invertebrates, and soil heterotrophic processes, LOECs were used as screening benchmarks for effect levels in the environment. In most cases, too few data were available to derive a screening benchmark or to have a high degree of confidence in the benchmarks that were derived. In most cases, benchmarks for soil microbial processes could not be derived from degradation studies because the microorganisms were allowed to acclimate to the chemicals.

Acknowledgments. The development of screening benchmarks for munitions compounds involved the participation of many individuals. Robert H. Ross and Patricia H. Reno of the Chemical Hazard Evaluation Group, Life Sciences Division, ORNL, provided project guidance, chapter review, and editorial comments. Lee Ann Wilson and Kimberly G. Slusher conducted literature searches and retrieved documents. Glenn W. Suter II and Bradley E. Sample of the Environmental Sciences Division, ORNL, provided peer review for some of the chapters.

This research was performed in support of a USEPA program to develop biomarkers for quantifying the exposures of ecological receptors to nitroaromatic munitions compounds and the U.S. Army program to establish cleanup levels for munitions contaminants. The research was funded by the Department of Defense's Strategic Environmental Research and Development Program (SERDP) via funds provided to the USEPA (F.B. Daniel, Project Officer) and the U.S. Army (R.M. Muhly, Project Officer).

References

ACS (American Chemical Society) (1998) SciFinder Online Information Retrieval Service, American Chemical Society, Columbus, OH.

Alvarez MA, Kitts CL, Botsford JL, Unkefer PJ (1995) *Pseudomonas aeruginosa* strain MA01 aerobically metabolizes the aminodinitrotoluenes produced by 2,4,6-trinitrotoluene nitro group reduction. Can J Microbiol 41:984–991.

Andren RK, Nystron JM, McDonnell RP, Stevens BW (1977) Explosives removal from munitions wastewater. Proc Ind Waste Conf 30:816–825.

ASTM (1980) Standard practice for conducting acute toxicity tests with fishes, macroinvertebrates, and amphibians. E-729-80. American Society for Testing and Materials, Philadelphia, PA.

Atkinson R (1987) A structure-activity relationship for the estimation of rate constants for the gas-phase reactions of OH radicals with organic compounds. Int J Chem Kinet 19:799–828.

ATSDR (Agency for Toxic Substances and Disease Registry) (1994) Toxicological profile for HMX (draft). Public Health Service, U.S. Department of Health and Human Services, Washington, DC.

ATSDR (1995a) Toxicological profile for 2,4,6-trinitrotoluene. Public Health Service, U.S. Department of Health and Human Services, Washington, DC.

ATSDR (1995b) Toxicological profile for 1,3-dinitrobenzene and 1,3,5-trinitrobenzene (draft). Public Health Service, U.S. Department of Health and Human Services, Washington, DC.

ATSDR (1995c) Toxicological profile for RDX. Public Health Service, U.S. Department of Health and Human Services, Washington, DC.

ATSDR (1995d) Toxicological profile for tetryl. Public Health Service, U.S. Department of Health and Human Services, Washington, DC.

Bailey HC (1982a) Development and testing of a laboratory model ecosystem for use in evaluating biological effects and chemical fate of pollutants. In: Pearson JG, Foster RB, Bishop WE (eds) Aquatic Toxicology and Hazard Assessment, 5th Conference. ASTM STP 766. American Society for Testing and Materials, Philadelphia, PA, pp 221–233.

Bailey HC (1982b) Toxicity of TNT wastewater to aquatic organisms. Monthly Prog Rep 80. SRI International, Menlo Park, CA. (Cited in van der Schalie 1983.)

Bailey HC, Spanggord RJ (1983) The relationship between the toxicity and structure of nitroaromatic chemicals. In: Bishop WE, Cardwell RD, Heidolph BB (eds) Aquatic Toxicology and Hazard Assessment, 6th Symposium. ASTM STP 802. American Society for Testing and Materials, Philadelphia, PA, pp 98–107.

Bailey HC, Spanggord RJ, Javitz HS, Liu DHW (1985) Toxicity of TNT wastewaters to aquatic organisms. Final Report. Vol. III. Chronic toxicity of LAP wastewater and 2,4,6-trinitrotoluene. AD-A164 282. SRI International, Menlo Park, CA.

Banerjee S, Yalkowsky SH, Valvant SC (1980) Water solubility and octanol/water partition coefficients of organics. Limitations of the solubility-partition coefficient correlation. Environ Sci Technol 14:1227–1230.

Banwart WL, Hassett JJ (1990) Effect of soil amendments on plant tolerance and extractable TNT from TNT contaminated soils. Agron Abstr 83:33.

Bauer JW (ed) (1985) Groundwater monitoring study No. 38-26-0457-86. AMC open burning/open detonation facilities, February 1984–March 1985. U.S. Army Materiel Command, Alexandria, VA.

Bel P, Ketcha MM, Pollard DL, Caldwell DJ, Martin JP, Narayanan L, Fisher JW (1994) *In vivo* metabolism of 1,3,5-trinitrobenzene in rats. In: 42nd Annual Conference of the American Society for Mass Spectrometry and Allied Topics, Chicago, IL, May 29–June 2, 1994.

Belkin F, Bishop RW, Sheely MV (1985) Analysis of explosives in water by capillary gas chromatography. J Chromatogr Sci 24:532–534.

Bender ES, Robinson PF, Moore MW, Thornton WD, Asaki AE (1977) Preliminary environmental survey of Holston Army Ammunition Plant, Kingsport, TN. AD-A043 662. U.S. Army Chemical Systems Laboratory, Aberdeen Proving Ground, MD.

Bentley RE, Dean JW, Ellis SJ, Hollister TA, LeBlanc GA, Sauter S, Sleight BH (1977a) Laboratory evaluation of the toxicity of cyclotrimethylene trinitramine (RDX) to aquatic organisms. AD A061730. Final Report. EG&G Bionomics, Wareham, MA, for U.S. Army Medical Bioengineering Research and Development Laboratory, Fort Detrick, MD.

Bentley RE, LeBlanc GA, Hollister TA, Sleight BH III (1977b) Acute toxicity of 1,3,5,7-tetranitro-octahydro-1,3,5,7-tetrazocine (HMX) to aquatic organisms. Final report. AD A061 730. EG & G Bionomics, Wareham, MA.

Bentley RE, Petrocelli SR, Suprenant DC (1984) Determination of the toxicity to aquatic organisms of HMX and related wastewater constituents. Part III. Toxicity of HMX, TAX and SEX to aquatic organisms. Final report. AD A172 385. Springborn Bionomics, Inc., Wareham, MD.

Benya TJ, Cornish HH (1994) Aromatic nitro and amino compounds. In: Patty's Industrial Hygiene and Toxicology, 4th Ed., Vol. IIB. Wiley, New York.

Bollog J-M, Loll MJ (1983) Incorporation of xenobiotics into soil humus. Experentia (Basel) 39:1221–1231.

Boopathy R, Manning J, Montemagno C, Rimkus K (1994) Metabolism of trinitrobenzene by a *Pseudomonas* consortium. Can J Microbiol 40:787–790.

Bringmann G, Kuhn R (1959) The toxic effect of waste water on aquatic bacteria, algae, and small crustaceans. Gesundh Ing 80:115–120. (Cited in Burrows and Dacre 1975.)

Bringmann G, Kuhn R (1978) Testing of substances for their toxicity threshold: model organisms *Microcystis (Diplocystis) aeruginosa* and *Scenedesmus quadricauda*. Mitt Int Ver Theor Angew Limnol 21:275–284.

Bringmann G, Kuhn R (1980) Comparison of the toxicity thresholds of water pollutants to bacteria, algae, and protozoa in the cell multiplication inhibition test. Water Res 14:231–241.

Budavari S, O'Neil MJ, Smith A, Heckelman PE (eds) (1996) The Merck Index, 12th Ed. Merck & Co., Rahway, NJ, pp 461, 553, 554, 1129, 1657, 1658.

Bumpus JA, Tatarko M (1994) Biodegradation of 2,4,6-trinitrotoluene by *Phanerochaete chrysosporium*: identification of initial degradation products and the discovery of a TNT metabolite that inhibits lignin peroxidases. Curr Microbiol 28:185–190.

Burlinson NE (1980) Fate of TNT in an aquatic environment: Photodecomposition vs.

biotransformation. NSWC TR 79-445, AD B045846. Naval Surface Weapons Center, White Oak, Silver Spring, MD.

Burlinson NE, Glover DJ (1977) Photochemistry of TNT and related nitrobodies. Quarterly Progress Report No. 14 for 1 October to 31 December 1977. Explosive Chemistry Branch, Naval Surface Weapons Center, White Oak, Silver Spring, MD.

Burlinson NE, Kaplan LA, Adams CE (1973) Photochemistry of TNT: investigation of the pink water problem. AD-769 670. National Technical Information Service, Springfield, VA.

Burrows D, Dacre JC (1975) Toxicity to aquatic organisms and chemistry of nine selected waterborne pollutants from munitions manufacture—a literature review. AD A010 660. U.S. Army Biomedical Research and Development Laboratory, Fort Detrick, MD.

Burrows WD, Chyrek RH, Noss CI (1984) Treatment for removal of munition chemicals from Army industrial wastewaters. In: Lagrega MD, Long DA (eds) (1984) Toxic and Hazardous Wastes: Proceedings of the 16th Mid-Atlantic Industrial Waste Conference, Bucknell University. Technomic Publishing, Lancaster, PA, pp 331–342.

Burrows EP, Rosenblatt DH, Mitchell WR, Parmer DL (1989) Organic explosives and related compounds: environmental and health considerations. AD-A210 554. U.S. Army Biomedical Research and Development Laboratory, Fort Detrick, MD.

Cameron BD (1986) HMX: toxicokinetics of ^{14}C-HMX following oral administration to the rat and mouse and intravenous administration to the rat. Final report. AD A171600. U.S. Army Medical Research and Development Command, Fort Detrick, MD.

Carpenter DF, McCormick NG, Cornell JH, Kaplan AM (1978) Microbial transformation of ^{14}C-labeled 2,4,6-trinitrotoluene in an activated sludge system. Appl Environ Microbiol 35:949.

Cataldo DA, Harvey SD, Fellows RJ (1993) The environmental behavior and chemical rate of energetic compounds (TNT, RDX, tetryl) in soil and plant systems. PNL-SA-22363. Presented at the 17th Annual Army Environmental R&D Symposium and 3rd USACE Innovative Technology Transfer Workshop, June 22–24, 1993, Williamsburg, VA.

Cataldo DA, Harvey SD, Fellows RJ, Bean RM, McVetty BD (1989) An evaluation of the environmental fate and behavior of munitions materiel (TNT, RDX) in soil and plant systems. PNL-7370, AD-A223 546. U.S. Army Medical Research and Development Command, Fort Detrick, MD.

Chandler CD, Kohlbeck JA, Bolleter WJ (1972) Continuous TNT process studies. III. Thin-layer chromatographic analysis of oxidation products from nitration. J Chromatogr 64:123–128. (Cited in Wentzel et al., 1979.)

Checkai RT, Major MA, Nwanguma RO, Amos JC, Phillips CT, Wentsel RS, Sadusky MC (1993) Transport and fate of nitroaromatic and nitramine explosives in soils from open burning/open detonation operations: Milan Army Ammunition Plant (MAAP). AD-A279 145, ERDEC-TR-136. Edgewood Research, Development & Engineering Center, Aberdeen Proving Ground, MD.

Cholakis JM, Wong LCK, Van Goethem DL, Minor J, Short R, Spring H, Ellis HV III (1980) Mammalian toxicological evaluation of RDX. Final report. AD A092531, DMD17-78-C-8027. Midwest Research Institute, Kansas City, MO.

Clear J, Collins P (1982) Final report for the Iowa Army Ammunition Plant (IAAP). Iowa Army Ammunition Plant, Middletown, IA.

Cody TE, Witherup S, Hastings L, Stemmer K, Christian RT (1981) 1,3-Dinitrobenzene: toxic effects *in vivo* and *in vitro*. J Toxicol Environ Health 7:829–848.

Cooper JR, Caldwell DJ (1995) Developmental toxicity evaluation of 1,3,5-trinitrobenzene in Sprague-Dawley rats. Final report, U.S. Army, Wright-Patterson AFB, OH. (cited in USEPA, 1977).

Curtis MW, Ward CH (1981) Aquatic toxicity of forty industrial chemicals: testing in support of hazardous substance spill prevention regulation. J Hydrol 51:359–367.

Daniele E (1964) Blood clotting alterations in chronic laboratory tetryl poisoning. Folia Med 8:767–776 (in Italian). (Cited in ATSDR 1995d.)

Deaver GH, Tucker RC, Adams W, McCann M (1986) Midwest site confirmatory survey assessment report for Joliet Army Ammunition Plant. AMXTH-IR-CR 86095. U.S. Army Toxic and Hazardous Materials Agency, Aberdeen Proving Ground, MD.

Deneer JW, Seinen W, Hermens, JLM (1988) Growth of *Daphnia magna* exposed to mixtures of chemicals with diverse modes of action. Ecotoxicol Environ Saf 15:72–77.

Deneer JW, Sinnige TL, Seinen W, Hermens JLM (1987) Quantitative structure-activity relationships for the toxicity and bioconcentration factor of nitrobenzene derivatives towards the guppy (*Poecilia reticulata*). Aquat Toxicol 10:115–127.

Deneer JW, van Leeuwen CJ, Seinen W (1989) QSAR study of the toxicity of nitrobenzene derivatives towards *Daphnia magna, Chlorella pyrenoidosa* and *Photobacterium phosphoreum*. Aquat Toxicol 15:83–98.

Dey S, Godbole SH (1986) Biotransformation of *m*-dinitrobenzene by *Candida pulcherrima*. Indian J Exp Biol 24:29–33.

Dey S, Kanekar P, Godbole SH (1986) Aerobic degradation of *m*-dinitrobenzene. Indian J Environ Health 29:118–128.

Di Toro DM (1985) A particle interaction model of reversible organic chemical sorption. Chemosphere 14:1503–1538.

Di Toro DM, Zarba CS, Hansen DJ, Berry WJ, Swartz RC, Cowan CE, Pavlou SP, Allen HE, Thomas NA, Paquin PR (1991) Technical basis for establishing sediment quality criteria for nonionic organic chemicals using equilibrium partitioning. Environ Toxicol Chem 10:1541–1583.

Dilley JV, Tyson CA, Spanggord RJ, Sasmore DP, Newell GW, Dacre JC (1982) Short-term oral toxicity of 2,4,6-trinitrotoluene in mice, rats and dogs. J Toxicol Environ Health 9:565–585.

Drzyzga O, Gorontzy T, Schmidt A, Blotevogel KH (1995) Toxicity of explosives and related compounds to the luminescent bacterium *Vibrio fischeri* NRRL-B-11177. Arch Environ Contam Toxicol 28:229–235.

DuBois FW, Baytos JF (1991) Weathering of explosives for twenty years. LA-11931. Los Alamos National Laboratory, Los Alamos, NM.

El-hawari AM, Hodgson JR, Winston JM, Sawyer JM, Hainje M, Lee C-C (1981) Species differences in the disposition and metabolism of 2,4,6-trinitrotoluene as a function of route of administration. Final report. AD A114025. U.S. Army Medical Research and Development Command, Fort Detrick, MD.

Envirodyne Engineers, Inc. (1980) Milan Army Ammunition Plant contamination survey. Final report. AD-BO53362. Envirodyne Engineers, Inc., St. Louis, MO.

ESE (Environmental Science and Engineering, Inc) (1986) Alabama Army Ammunition Plant remedial investigation final report. U.S. Army Toxic and Hazardous Materials Agency, Installation Restoration Division, Aberdeen Proving Ground, MD.

Etnier EL (1986) Water quality criteria for hexahydro-1,3,5-trinitro-1,3,5-triazine (RDX).

ORNL-6178. Oak Ridge National Laboratory, Oak Ridge, TN. U.S. Army Medical Research and Development Command, Fort Detrick, MD.

Everett DJ, Maddock SM (1985) HMX: 13-week toxicity study in mice by dietary administration. AD A171602. Final report to the U.S. Army. Inveresk Research International, Ltd., Musselburgh, Scotland.

Everett DJ, Johnson IR, Hudson P, Jones M (1985) HMX: 13-week toxicity study in rats by dietary administration. AD A171601. Final report to the U.S. Army. Inveresk Research International, Ltd., Musselburgh, Scotland.

Fati S, Daniele E (1965) Histopathological findings in chronic laboratory poisoning. Folia Med 1:269–276 (in Italian). (Cited in ATSDR 1995d.)

Fedoroff BT, Sheffield OE (1966) Cyclotrimethylenetrinitramine, cyclonite or RDX. In: Encyclopedia of Explosives and Related Items. Vol. 3. Picatinny Arsenal, Dover, NJ.

Fedoroff BT, Sheffield OE, Reese EF, Clift GD (1962) Encyclopedia of Explosives and Related Items. PATR 2700. Vol. 2. Picatinny Arsenal, Dover, NJ.

Fedoroff BT, Reese EF, Aaronson HA, Sheffield OE, Clift GD (1960) Encyclopedia of Explosives and Related Items. Vol. 1. Picatinny Arsenal, Dover, NJ.

Fellows RJ, Harvey SD, Cataldo DA (1992) An evaluation of the environmental fate and behavior of munitions materiel (tetryl and polar metabolites of TNT) in soil and plant systems. AD-A266 548. U.S. Army Medical Research and Development Command, Fort Detrick, MD.

Fernando T, Bumpus JA, Aust SD (1990) Biodegradation of TNT (2,4,6-trinitrotoluene) by *Phanerochaete chrysosporium*. Appl Environ Microbiol 56:1666–1671.

Fitzgerald GP, Gerlogg GC, Skong F (1952) Studies on chemicals with selective toxicity to bluegreen algae. Sewage Ind Wastes 24:888–896.

Freeman DJ, Colitti OA (1982) Removal of explosives from load-assemble-pack wastewater (pink water) using surfactant technology. Proc Ind Waste Conf 36:383–394.

Funk SB, Roberts DJ, Crawford DL, Crawford RL (1993) Initial-phase optimization for bioremediation of munition compound-contaminated soils. Appl Environ Microbiol 59:2171–2177.

Furedi EM, Levine BS, Sagartz JW, Rac VS, Lish PM (1984a) Determination of the chronic mammalian toxicological effects of TNT: twenty-four month chronic toxicity/carcinogenicity study of trinitrotoluene (TNT) in the Fischer-344 rat. Final report, phase III. Vol. I. AD-A168 637. U.S. Army Medical Research and Development Command, Fort Detrick, MD. IIT Research Institute, Chicago, IL.

Furedi EM, Levine BS, Sagartz JW, Rac VS, Lish PM (1984b) Determination of the chronic mammalian toxicological effects of TNT: twenty-four month chronic toxicity/carcinogenicity study of trinitrotoluene (TNT) in the B6C3F1 hybrid mouse. Final report, phase IV. Vol. I. AD-A168 754. U.S. Army Medical Research and Development Command, Fort Detrick, MD. IIT Research Institute, Chicago, IL.

Goerlitz DF (1992) A review of studies of contaminated groundwater conducted by the U.S. Geological Survey Organics Project, Menlo Park, California, 1961–1990. Environ Sci Pollut Control Ser 4:295–355.

Goerlitz DF, Franks BJ (1989) Use of on-site high performance liquid chromatography to evaluate the magnitude and extent of organic contaminants in aquifers. Ground Water Monit Rev 9:122–129.

Grant CL, Jenkins TF, Myers KF, McCormick EF (1995) Holding-time estimates for soils containing explosives residues: comparison of fortification vs. field contamination. Environ Toxicol Chem 14:1865–1874.

Greene B, Kaplan DL, Kaplan AM (1984) Degradation of pink water compounds in soil—TNT, RDX, HMX. AD-A157 954. U.S. Army Natick Research and Development Center, Natick, MA.
Gregory RG, Elliott WG (1987) Remedial Investigation at Louisiana Army Ammunition Plant. Final report. AMXTH-IR-CR-87110. Environmental and Science Engineering, Inc., Gainsville, FL. U.S. Army Toxic and Hazardous Materials Agency, Aberdeen Proving Ground, MD.
Grindley J (1946) Toxicity to rainbow trout and minnows of some substances known to be present in waste water discharged to rivers. Ann Appl Biol 33:103–112.
Haderlein SB, Schwarzenbach RP (1993) Adsorption of substituted nitrobenzenes and nitrophenols to mineral surfaces. Environ Sci Technol 27:316–326.
Hale VQ, Stanford TB, Taft LG (1979) Evaluation of the environmental fate of munitions compounds in soil. AD A034226. U.S. Army Medical Bioengineering Research and Development Laboratory, Fort Detrick, MD. Battelle Columbus Laboratories, Columbus, OH.
Hallas LE, Alexander M (1983) Microbial transformation of nitroaromatic compounds in sewage effluent. Appl Environ Microbiol 45:1234–1241.
Hansch C, Leo A (1979) Substituent Constants for Correlation Analysis in Chemistry and Biology. Wiley, New York.
Hansch C, Leo AJ (1985) Medchem Project, Issue No. 25, Claremont, CA.
Haroun LA, MacDonell MM, Peterson JM (1990) Multimedia assessment of health risks for the Weldon Spring site remedial action project. Proc AWMA Annu Meet 83:19.
Hart ER (1976) Two-year feeding study in rats. Final report. AD A040161. Litton Bionetics, Inc., Kensington, MD. (N00014-73-C-0162, NR202-043.)
Harvey SD, Fellows RJ, Cataldo DA, Bean RM (1990) Analysis of 2,4,6-trinitrotoluene and its transformation products in soils and plant tissues by high-performance liquid chromatography. J Chromatogr 518:361–374.
Harvey SD, Fellows RJ, Cataldo DA, Bean RM (1991) Fate of the explosive hexahydro-1,3,5-trinitro-1,3,5-triazine (RDX) in soil and bioaccumulation in bush bean hydroponic plants. Environ Toxicol Chem 10:845–855.
Harvey SD, Fellows RJ, Campbell JA, Cataldo DA (1992) Analysis of the explosive 2,4,6-trinitrophenylmethylnitramine (tetryl) and its transformation products in soil. J Chromatogr 605:227–240.
Harvey SD, Fellows RJ, Cataldo DA, Bean RM (1993) Analysis of the explosive 2,4,6-trinitrophenylmethylnitramine (tetryl) in bush bean plants. J Chromatogr 630:167–177.
Hermens J, Canton H, Steyger N, Wegman R (1984) Joint effects of a mixture of 14 chemicals on mortality and inhibition of reproduction of *Daphnia magna*. Aquatic Toxicol 5:315–322.
Hodgson JR, Winston JM, House WB, El-Hawari AM, Murrill EE, Weigand WJ, Burton W, Lee C-C (1977) Evaluation of difference in mammalian metabolism of trinitrotoluene (TNT) as a function of route of administration and carcinogenesis testing. Annual progress report No. 1. AD BO24821L. U.S. Army Medical Research and Development Command, Washington, DC.
Hoffsommer JC, Rosen JM (1972) Analysis of explosives in sea water. Bull Environ Contam Toxicol 7:177–181.
Hoffsommer JC, Rosen JM (1973) Hydrolysis of explosives in seawater. Bull Environ Contam Toxicol 10:78–79.
Hoffsommer JC, Glover DJ, Rosen JM (1972) Analysis of explosives in sea water and

in ocean floor sediment and fauna. AD 757778, NOLTR-72-215. Naval Ordinance Laboratory, White Oak, Silver Springs, MD.

Hoffsommer JC, Kaplan LA, Glover DJ, Kubose DA, Dickinson C, Goya H, Kayser EG, Groves CL, Sitzmann ME (1978) Biodegradability of TNT: a three-year pilot plant study. Final report. NSWC/WOL TR77-136, AD A061144. Naval Surface Weapons Center, White Oak, Silver Springs, MD.

Howard PH, Jarvis WF, Sage GW (1989) Handbook of environmental fate and exposure data for organic chemicals. Lewis Publishers, Chelsea, MI.

Howard PH, Boethling RS, Davis WF (1991) Handbook of environmental degradation rates. Lewis Publishers, Chelsea, MI, pp 454–455. (Cited in ATSDR 1995a.)

Howard PH, Santodonato J, Saxena J, Malling J, Greninger D (1976) Investigation of selected potential environmental contaminants: Nitroaromatics. EPA/560/2-76-010. USEPA Office of Toxic Substances, Washington, DC.

HSDB (1995a) Hazardous Substances Data Bank. 2,4,6-Trinitrotoluene. MEDLARS Online Information Retrieval System, National Library of Medicine, Bethesda, MD.

HSDB (1995b) Hazardous Substances Data Bank. 1,3,5-Trinitrobenzene. MEDLARS Online Information Retrieval System, National Library of Medicine, Bethesda, MD.

HSDB (1995c) Hazardous Substances Data Bank. 1,3-Drinitrobenzene. MEDLARS Online Information Retrieval System, National Library of Medicine, Bethesda, MD.

HSDB (1995d) Hazardous Substances Data Bank. 2,4-Dinitroaniline. MEDLARS Online Information Retrieval System, National Library of Medicine, Bethesda, MD.

HSDB (1995e) Hazardous Substances Data Bank. RDX. MEDLARS Online Information Retrieval System, National Library of Medicine, Bethesda, MD.

HSDB (1995f) Hazardous Substances Data Bank. HMX. MEDLARS Online Information Retrieval System, National Library of Medicine, Bethesda, MD.

HSDB (1995g) Hazardous Substances Data Bank. Tetryl. MEDLARS Online Information Retrieval System, National Library of Medicine, Bethesda, MD.

Jenkins TF (1989) Development of an analytical method for the determination of extractable nitroaromatics and nitramines in soils. Ph.D. thesis. University of New Hampshire, Durham, NH.

Jenkins TF, Grant CL (1987) Comparison of extraction techniques for munitions residues in soil. Anal Chem 59:1326–1331.

Jenkins TF, Leggett DC, Grant CL, Bauer CF (1986) Reversed-phase high-performance liquid chromatographic determination of nitroorganics in munitions wastewater. Anal Chem 58:170–175.

Jerger DE, Simon PB, Weitzel RL, Schenk JE (1976) Aquatic field surveys at Iowa, Radford, and Joliet Army Ammunition Plants. Vol. III. Microbiological investigations, Iowa and Joliet Army Ammunition Plants. AD AO36778. U.S. Army Medical Research and Development Command, Washington, DC.

Jones DS, Hull RN, Suter GW II (1996) Toxicological benchmarks for screening contaminants of potential concern for effects on sediment-associated biota: 1996 revision. ES/ER/TM-95/R2. Oak Ridge National Laboratory, Oak Ridge, TN.

Kaye SM (1980a) The dinitro ortho- meta- and para-toluidines. In: Encyclopedia of Explosives and Related Items. Vol. 9. U.S. Army Armament Research and Development Command, Large Caliber Weapons Systems Laboratory, Dover, NJ.

Kaye SM (1980b) Tetryl. In: Encyclopedia of Explosives and Related Items. Vol. 9. U.S. Army Armament Research and Development Command, Large Caliber Weapons Systems Laboratory, Dover, NJ.

Kayser EG, Burlinson NE (1988) Migration of explosives in soil: analysis of RDX, TNT, and tetryl from a ^{14}C lysimeter study. J Energy Mater 6:45–71.

Kayser EG, Burlinson NE, Rosenblatt DH (1984) Kinetics of hydrolysis and products of hydrolysis and photolysis of tetryl. AD-A153144, NSWC TR 84-68. Naval Surface Weapons Center (Code R16), White Oak, Silver Spring, MD.

Kinkead ER, Wolf RE, Flemming CD, Caldwell DJ, Miller CR, Marit GB (1995) Reproductive toxicity screen of 1,3,5-trinitrobenzene administered in the diet of Sprague-Dawley rats. Toxicol Ind Health 11:309–323.

Kitchens JF, Harward WE, Lauter DM, Wentsel RS, Valentine RS (1978) Preliminary problem definition study of 48 munitions-related chemicals. I. Explosives related chemicals. Final report. AD 066 307. Atlantic Research Corporation, Alexandria, VA.

Klausmeier RE, Osmon JL, Walls DR (1973) The effect of trinitrotoluene on microorganisms. Dev Ind Microbiol 15:309–317.

Kraus DL, Hendry CD, Keirn MA (1985) U.S. Department of Defense superfund implementation at a former TNT manufacturing facility. In: 6th National Conference on Management of Uncontrolled Hazardous Waste Sites, Washington, DC, November 4–6, 1985. Hazardous Materials Control Research Institute, pp 314–318. (Cited in ATSDR 1993.)

Kubose DA, Hoffsommer JC (1977) Photolysis of RDX in aqueous solution. Initial studies. AD A-42199/OST. NTIS, Springfield, VA.

Lakings DB, Gan O (1981) Identification or development of chemical analysis methods for plants and animal tissues. AD-A103085. Midwest Research Institute, Kansas City, MO. U.S. Army, Fort Detrick, MD. (Cited in Layton et al. 1987.)

Layton D, Mallon B, Mitchell W, Hall L, Fish R, Perry L. Snyder G, Bogen K, Malloch W, Ham C, Dowd P (1987) Conventional weapons demilitarization: a health and environmental effects data base assessment. Explosives and their co-contaminants. Final report, phase II. AD-A220588. Lawrence Livermore National Laboratory, Livermore, CA. U.S. Army Medical Research and Development Command, Frederick, MD.

Levine BS, Furedi EM, Gordon DE (1983) Determination of the chronic mammalian toxicological effects of RDX: twenty-four month chronic toxicity/carcinogenicity study of hexahydro-1,3,5-trinitro-1,3,5-triazine (RDX) in the Fischer-344 rat. Phase V, final report. Vol. I. ADA 160 774. IIT Research Institute, Chicago, IL. U.S. Army Medical Research and Development Command, Fort Detrick, Frederick, MD.

Levine BS, Furedi EM, Gordon DE, Burns JM, Lish PM (1981) Thirteen week oral (diet) toxicity study of trinitrotoluene (TNT), hexahydro-1,3,5-trinitro-1,3,5-triazine (RDX) and TNT/RDX mixtures in the Fischer-344 rat. AD A108 447. U.S. Army Medical Research and Development Command, Fort Detrick, MD. IIT Research Institute, Chicago, IL.

Levine BS, Furedi EM, Gordon DE, Lish PM, Barkley JJ (1984) Subchronic toxicity of trinitrotoluene in Fischer-344 rats. Toxicology 32:253–265.

Levine BS, Rust JH, Barkley JJ, Furedi EM, Lish PM (1990) Six month oral toxicity study of trinitrotoluene in beagle dogs. Toxicology 63:233–244.

Linder RE, Hess RA, Strader LF (1986) Testicular toxicity and infertility in male rats treated with 1,3-dinitrobenzene. J Toxicol Environ Health 19:477–489.

Lindner V (1978) Explosives. In: Kirk-Othmer Encyclopedia of Chemical Technology, 3rd Ed. Vol. 9. Wiley, New York, pp 581–584.

Lish PM, Levine BS, Furedi EM, Sagartz EM, Rac VS (1984) Determination of the chronic mammalian toxicological effects of RDX: twenty-four month chronic toxicity/carcinogenicity study of hexahydro-1,3,5-trinitro-1,3,5-triazine (RDX) in the

B6C3F1 hybrid mouse. Phase VI. Vol. 1. AD A160774. IIT Research Institute, Chicago, IL. U.S. Army Medical Research and Development Command, Frederick, MD.
Liu DH, Bailey HC, Pearson JG (1983a) Toxicity of a complex munitions wastewater to aquatic organisms. In: Bishop WE, Cardwell RD, Heidolph BB (eds) Aquatic Toxicology and Hazard Assessment: Sixth Symposium. ASTM STP 802. American Society for Testing and Materials, Philadelphia, PA, pp 135–150.
Liu DH, Spanggord RJ, Bailey HC, Javitz HS, Jones DCL (1983b) Toxicity of TNT wastewaters to aquatic organisms. Final report. Vol. I: Acute toxicity of LAP wastewater and 2,4,6-trinitrotoluene. AD A142 144. SRI International, Menlo Park, CA.
Lyman WJ, Reehl WF, Rosenblatt DH (1982) Handbook of Chemical Property Estimation Methods: Environmental Behavior of Organic Compounds. McGraw-Hill, New York.
Mabey WR, Tse D, Baraze A, Mill T (1983) Photolysis of nitroaromatics in aquatic systems. I. 2,4,6-Trinitrotoluene. Chemosphere 12:3–16.
Marvin-Sikkema FD, de Bout JAM (1994) Degradation of nitroaromatic compounds by microorganisms. Appl Microbiol Biotechnol 42:499–507.
Matthews HB, Chopade HM, Smith RW, Burka LT (1986) Disposition of 2,4-dinitroaniline in the male F-344 rat. Xenobiotica 16:1–10.
McCormick NG, Feeherry FE, Levinson HS (1976) Microbial transformation of 2,4,6-trinitrotoluene and other nitroaromatic compounds. Appl Environ Microbiol 31:949–958.
McCormick NG, Cornell JH, Kaplan AM (1981) Biodegradation of hexahydro-1,3,5-trinitro-1,3,5-triazine. Appl Environ Microbiol 42:817–823.
McCormick NG, Cornell JH, Kaplan AM (1984) The anaerobic biotransformation of RDX, HMX, and their acetylated derivatives. AD-A149 464. U.S. Army Natick Research and Development Center, Natick, MA. U.S. Army Toxic and Hazardous Materials Agency, Aberdeen Proving Ground, MD.
McEuen SF, Jacobson CF, Brown CD, Miller MG (1995) Metabolism and testicular toxicity of 1,3-dinitrobenzene in the rat: effect of route of administration. Fundam Appl Toxicol 28:94–99.
McFarlane C, Nolt C, Wickliff C, Pfleeger T, Shimabuku R, McDowell M (1987) The uptake, distribution, and metabolism of four organic chemicals by soybean plants and barley roots. Environ Toxicol Chem 6:847–856.
McKone TE, Layton DW (1986) Screening the potential risks of toxic substances using a multimedia compartment model: estimation of human exposure. Regul Toxicol Pharmacol 6:359–380.
McLeese DW, Zitko V, Peterson MR (1979) Structure-lethality relationships for phenols, anilines and other aromatic compounds in shrimp and clams. Chemosphere 8:53–57.
McLellan WL, Hartley WR, Bower ME (1988) Octahydro-1,3,5,7-tetranitro-1,3,5,7-tetrazocine (HMX). In: Roberts WC, Hartley WR (eds) Drinking Water Health Advisory: Munitions. Office of Drinking Water Health Advisories, U.S. Environmental Protection Agency, Washington, DC, pp 247–273.
McNamara BP, Averill HP, Owens EJ (1974) The toxicology of cyclotrimethylenetrinitramine (RDX) and cyclotetra-methylenetetranitramine (HMX) solutions in dimethylsulfoxide (DMSO), cyclohexanone, and acetone. AD-780010. U.S. Army, Aberdeen Proving Ground, MD.
Mitchell WR, Dennis WH (1982) Biodegradation of 1,3-dinitrobenzene. J Environ Sci Health A17:837–853.
Mitchell WR, Dennis WH, Burrows EP (1982) Microbial interaction with several munitions compounds: 1,3-dinitrobenzene, 1,3,5-trinitrobenzene, and 3,5-dinitroaniline. AD A116651. U.S. Army Medical Bioengineering Research and Development Laboratory, Fort Detrick, MD.

Monnot D, Kennedy D, Cira D, Starkey D (1982) Cornhusker Army Ammunition Plant. Final report. DRXTH-AS-CR-82155. Conducted by Envirodyne Engineers, Inc., St. Louis, MO, for Cornhusker Army Ammunition Plant, Grand Island, NE. U.S. Army Toxic and Hazardous Materials Agency, Aberdeen Proving Ground, MD.

Naumova RP, Selivanovskaya SY, Cherepneva IE (1988) Conversion of 2,4,6-trinitrotoluene under conditions of oxygen and nitrate respiration of *Pseudomonas fluoresens*. Prikl Biokhim Mikrobiol 24:493–498.

Nay MW, Randall CW, King PH (1972) Factors affecting color development during treatment of TNT waste. Ind Wastes 18:20–29.

Nay MW, Randall CW, King PH (1974) Biological treatability of trinitrotoluene manufacturing wastewater. J Water Pollut Control Fed 46:485–497.

Newell EL Jr (1984) Phase 3. Hazardous waste study No. 37-26-0147-84. Summary of AMC open burning/open detonation ground evaluations, November 1981–September 1983. Department of the Army, U.S. Army Environmental Hygiene Agency, Aberdeen Proving Ground, MD.

Nystrom DD, Rickert DE (1987) Metabolism and excretion of dinitrobenzenes by male Fischer-344 rats. Drug Metab Dispos 15:821–825.

Opresko DM (1995a) Toxicity summary for 2,4,6-trinitrotoluene. Prepared by Oak Ridge National Laboratory for the U.S. Army Toxic and Hazardous Materials Agency, Aberdeen Proving Ground, MD.

Opresko DM (1995b) Review of biomonitoring studies and ecological surveys conducted at U.S. military installations. ORNL/M-4097. Oak Ridge National Laboratory, Oak Ridge, TN.

Osmon JL, Klausmeier RE (1972) The microbial degradation of explosives. Dev Ind Microbiol 14:321–325.

Palazzo AJ, Leggett DC (1986) Effect and disposition of TNT in a terrestrial plant. J Environ Qual 15:49–52.

Parke D (1961) Studies in detoxication. V. The metabolism of *m*-dinitro[C^{14}C]benzene in the rabbit. Biochem J 78:262–271.

Parmeggiani L, Bartalini E, Sassi C, Perini A (1956) Tetryl occupational diseases: experimental investigations and prevention. Med Lav 47:293–313 (in Italian). (Cited in ATSDR 1995.)

Parmelee RW, Wentsel RS, Phillips CT, Simini M, Checkai RT (1993) Soil microcosm for testing the effects of chemical pollutants on soil fauna communities and trophic structure. Environ Toxicol Chem 12:1477–1486.

Parrish FW (1977) Fungal transformation of 2,4-dinitrotoluene and 2,4,6-trinitrotoluene. Appl Environ Microbiol 34:232–233.

Pasti-Grigsby MB, Lewis TA, Crawford DL, Crawford RL (1996) Transformation of 2,4,6-trinitrotoluene (TNT) by actinomycetes isolated from TNT-contaminated and uncontaminated environments. Appl Environ Microbiol 62:1120–1123.

Pathology Associates, Inc. (1994) TNB toxicity evaluation in *Peromyscus* mice—90 day exposure. Study no. 94-105. Prepared under contract to USEPA, Environmental Systems Monitoring Laboratory, Cincinnati, OH, for U.S. Army Medical Research, Development, Acquisition and Logistics Command (Provisional), Fort Detrick, MD.

Patterson JW, Shapira NI, Brown J (1977) Pollution abatement in the military explosives industry. Proc Ind Waste Conf 31:385–394. (Cited in ATSDR 1995a.)

Pearson JG, Glennon JP, Barkley JJ, Highfill JW (1979) An approach to the toxicological evaluation of a complex industrial wastewater. In: Marking LL, Kimerle RA (eds) Aquatic Toxicology. ASTM STP 667. American Society for Testing and Materials, Philadelphia, PA, pp 284–301.

Pederson GL (1970) Evaluation of toxicity of selected TNT wastes on fish. Phase I: Acute toxicity of alpha-TNT to bluegills. Sanitary engineering special study no. 24-007-70/71. AD 725572. U.S. Army Environmental Hygiene Agency, Edgewood Arsenal, MD.

Pennington JC (1988a) Soil sorption and plant uptake of 2,4,6-trinitrotoluene. EL-88-12, AD A200 502. U.S. Army Biomedical Research and Development Laboratory, Fort Detrick, MD.

Pennington JC (1988b) Plant uptake of 2,4,6-trinitrotoluene, 4-amino-2,6-dinitrotouuene, and 2-amino-4,6-dinitrotoluene using ^{14}C-labeled and unlabeled compounds. EL-88-10, AD-A203 690. U.S. Army Biomedical Research and Development Laboratory, Fort Detrick, MD.

Pennington JC, Patrick WH Jr (1990) Adsorption and desorption of 2,4,6-trinitrotoluene by soils. J Environ Qual 19:559–567.

Pereira WE, Short DI, Manigold DB, Roscio PK (1979) Isolation and characterization of TNT and its metabolites in groundwater by gas chromatograph-mass spectrometer-computer techniques. Bull Environ Contam Toxicol 21:554–562.

Philbert MA, Gray AJ, Conners TA (1987) Preliminary investigations into the involvement of the intestinal microflora in CNS toxicity induced by 1,3-dinitrobenzene in male F-344 rats. Toxicol Lett 38:307–314.

Phillips CT, Checkai RT, Wentsel RS (1993) Toxicity of selected munitions and munition-contaminated soil on the earthworm (*Eisenia foetida*). AD-A264 408. Edgewood Research Development & Engineering Center, Aberdeen Proving Ground, MD.

Phillips CT, Checkai RT, Chester NA (1994) Toxicity testing of soil samples from Joliet Army Ammunition Plant, IL. AD-A279 091. Edgewood Research Development and Engineering Center, Aberdeen Proving Ground, MD.

Phung HT, Bulot MW (1981) Subsurface investigation of metal sludge and explosive disposal pond areas. In: Conway RA, Malloy DC (eds) Hazardous Solid Waste Testing, First Conference. ASTM STP 760. American Society for Testing and Materials, Philadelphia, PA.

Preuss A, Fimpel J, Diekert G (1993) Anaerobic transformation of 2,4,6-trinitrotoluene (TNT). Arch Microbiol 159:345–353.

Reddy G, Hampton AEG, Amos J, Major M (1996) Metabolism of 1,3,5-trinitrobenzene (TNB) *in vitro*. In: Annual Meeting of the Society for Toxicology, March 10–14, Anaheim, CA.

Reddy G, Reddy TV, Choudhury H, Daniel FB, Leach G (1997) Assessment of environmental hazards of 1,3,5-trinitrobenzene (TNB). J Toxicol Environ Health 52:447–460.

Reddy TV, Daniel FB, Robinson M, Olson GR, Wiechman B, Reddy G (1994a) Subchronic toxicity studies on 1,3,5-trinitrobenzene, 1,3-dinitrobenzene, and tetryl in rats: subchronic toxicity evaluation of 1,3,5-trinitrobenzene in Fischer-344 Rats. AD-A283 663/3/HDM. Prepared by USEPA, Environmental Monitoring Systems Laboratory, Cincinnati, OH, for the the U.S. Army Medical Research, Development, Acquisition and Logistics Command (Provisional), Fort Detrick, MD. National Technical Information Service, Springfield, VA.

Reddy TV, Daniel FB, Robinson M (1994b) Subchronic toxicity studies on 1,3,5-trinitrobenzene, 1,3-dinitrobenzene, and tetryl in rats: 14-day toxicity evaluation of *N*-methyl-*N*,2,4,6-tetranitroaniline in Fischer-344 rats. U.S. Environmental Protection Agency, Environmental Monitoring Systems Laboratory, Cincinnati, OH. AD-A284 190/6/HDM. National Technical Information System, Springfield, VA.

Reddy TV, Daniel FB, Robinson M, Olson GR, Weichman B, Reddy G (1994c) Subchronic toxicity studies on 1,3,5-trinitrobenzene, 1,3-dinitrobenzene, and tetryl in rats: subchronic toxicity evaluation of *N*-methyl-*N*-2,4,6-tetranitroaniline (tetryl) in Fischer-344 rats. U.S. Environmental Protection Agency, Environmental Monitoring Systems Laboratory, Cincinnati, OH. National Technical Information Service, Springfield, VA.

Reddy TV, Daniel FB, Robinson M (1995) Subchronic toxicity studies on 1,3,5-trinitrobenzene, 1,3-dinitrobenzene, and tetryl in rats: 90-day toxicity evaluation of 1,3-dinitrobenzene (DNB) in Fischer-344 rats. U.S. Environmental Protection Agency, Environmental Monitoring Systems Laboratory, Cincinnati, OH.

Reddy TV, Daniel FB, Olson GR, Wiechman BH, Reddy G (1996) Chronic toxicity studies on 1,3,5-trinitrobenzene in Fischer-344 rats. USEPA, Cincinnati, OH. ADA315216, U.S. Army Medical Research and Materiel Command, Frederick, MD.

Richards JJ, Junk GA (1986) Determination of munitions in water using macroreticular resins. Anal Chem 58:723–725.

Rosenblatt DH (1981) Environmental risk assessment for four munitions-related contaminants at Savannah army depot activity. Tech Rep 8110. U.S. Army Medical Bioengineering Research and Development Laboratory, Fort Detrick, MD.

Rosenblatt DH (1986) Contaminated soil cleanup objectives for Cornhusker Army Ammunition Plant. Tech Rep 8603. U.S. Army Medical Bioengineering Research and Development Laboratory, Fort Detrick, MD.

Rosenblatt DH, Small MJ (1981) Preliminary pollutant limit values for Alabama Army Ammunition Plant. AD-A104203. U.S. Army Medical Bioengineering Research and Development Laboratory, Fort Detrick, MD.

RTECS (1995) 2-Amino-4,6-dinitrotoluene. Registry of Toxic Effects of Chemical Substances, MEDLARS Online Information Retrieval System, National Library of Medicine, Washington, DC.

Ryon MG (1987) Water quality criteria for 2,4,6-trinitrotoluene (TNT). ORNL-6304. Oak Ridge National Laboratory, Oak Ridge, TN.

Ryon MG, Pal BC, Talmage SS, Ross RH (1984) Database assessment of the health and environmental effects of munition production waste products. ORNL-6018, AD-A145417. U.S. Army Medical Research and Development Command, Fort Detrick, MD.

Sample BE, Opresko DM, Suter GW II (1996) Toxicological benchmarks for wildlife: 1996 revision. ES/ER/TM-86/R3. Oak Ridge National Laboratory, Oak Ridge, TN.

Sanocki SL, Simon PB, Weitzel RL, Jerger DE, Schenk JE (1976) Aquatic field surveys at Iowa, Radford and Joliet Army Munition Plants. Final report. Vol. 1. Iowa Army Ammunition Plant. AD A036776. Environmental Control Technology Corporation, Ann Arbor, MI.

Schafer EW (1972) The acute oral toxicity of 369 pesticidal, pharmaceutical and other chemicals to wild birds. Toxicol Appl Pharmacol 21:315–330.

Schneider NR, Bradley SL, Andersen ME (1977) Toxicology of cyclotrimethylene-trinitramine: distribution and metabolism in miniature swine. Toxicol Appl Pharmacol 39: 531–541.

Schneider NR, Bradley SL, Andersen ME (1978) The distribution and metabolism of cyclotrimethylenetrinitramine (RDX) in the rat after subchronic administration. Toxicol Appl Pharmacol 46:163–171.

Schott CD, Worthley EG (1974) The toxicity of TNT and related wastes to an aquatic flowering plant, *Lemna perpusilla* Torr. AD-778 158. Aberdeen Proving Ground, MD.

Shugart LR, Griest WH, Tan E, Guzman Z, Caton JE, Ho CH, Tomkins BA (1990) TNT metabolites in animal tissues. Final report. ORNL/M-1336. Oak Ridge National Laboratory, Oak Ridge, TN.

Sikka HC, Banerjee S, Pack EJ, Appleton HT (1980) Environmental fate of RDX and TNT. U.S. Army Medical Research and Development Command, Fort Detrick, MD. Contract no. DAMD17-77-C-7026. Syracuse Research Corporation, Syracuse, NY.

Simini M, Rowland R, Lee EH, Wentsel RW (1992) Detection of stress in *Cucumis sativus* exposed to RDX using chlorophyll fluorescence. Abstract, 13th Annual Meeting, Society of Environmental Toxicology and Chemistry, Cincinnati, OH, November 8–12, p 48.

Simini M, Wentsel RS, Checkai RT, Phillips CT, Chester NA, Major MA, Amos JC (1995) Evaluation of soil toxicity at Joliet Army Ammunition Plant. Environ Toxicol Chem 14:623–630.

Simmons MS, Zepp RG (1986) Influence of humic substances on photolysis of nitroaromatic compounds in aqueous systems. Water Res 7:899–904.

Simon WW, Blackman GE (1953) Studies in the principles of phytotoxicity. IV. The effects of the degree of nitration on the toxicity of phenol and other substituted benzenes. J Exp Bot 4:235–250.

Small MJ, Rosenblatt DH (1974) Munitions production products of potential concern as waterborne pollutants: Phase II. AD-919031. U.S. Army Medical Bioengineering Research and Development Laboratory, Fort Detrick, MD.

Smith JH, Bomberger DC Jr, Haynes DL (1981) Volatilization rates of intermediate and low volatility chemicals from water. Chemosphere 10:281–291.

Smock LA, Stoneburner DL, Clark JR (1976) The toxic effects of trinitrotoluene (TNT) and its primary degradation products on two species of algae and the fathead minnow. Water Res 10:537–543.

Snell TW, Moffat BD (1992) A 2-d life cycle test with the rotifer *Brachionus calyciflorus*. Environ Toxicol Chem 11:1249–1257.

Soli G (1973) Microbial degradation of cyclonite (RDX). AD-762 751. Tech Publ 5525. Naval Weapons Center, China Lake, CA.

Spain JC (ed) (1995) Biodegradation of Nitroaromatic Compounds. Plenum Press, New York.

Spalding RF, Fulton JW (1988) Groundwater munition residues and nitrate near Grand Island, Nebraska, U.S.A. J Contam Hydrol 2:139–153.

Spanggord RJ, Gibson BW, Keck RG, Newell GW (1978) Mammalian toxicological evaluation of TNT wastewaters ("pink water"). Vol. I. Chemistry studies, draft report. AD A059434. U.S. Army Medical Research and Development Command, Fort Detrick, MD.

Spanggord RJ, Mill T, Chou TW, Mabey WR, Smith JH, Lee S (1980a) Environmental fate studies on certain munition wastewater constituents. Final report. Phase I: Literature review. SRI International, Menlo Park, CA. AD AO82372. U.S. Army Medical Research and Development Command, Fort Detrick, MD.

Spanggord RJ, Mabey WR, Mill T, Chou TW Smith JH, Lee S (1980b) Environmental fate studies on certain munition wastewater constituents. Final report. Phase II: Laboratory studies. SRI International, Menlo Park, CA. AD A099256. U.S. Army Medical Research and Development Command, Fort Detrick, MD.

Spanggord RJ, Mabey WR, Mill T, Chou T-W, Smith JH, Lee S (1981) Environmental fate studies on certain munition wastewater constituents. Phase III. Part I: Model validation. AD A129373. U.S. Army Medical Research and Development Command, Fort Detrick, MD.

Spanggord RJ, Gibson BW, Keck RG, Thomas DW (1982a) Effluent analysis of wastewater generated in the manufacture of 2,4,6-trinitrotoluene. 1. Characterization study. Environ Sci Technol 16:229–232.

Spanggord RJ, Mabey WR, Chou TW, Haynes DL, Alferness PL, Tee DS, Mill T (1982b) Environmental fate studies of HMX. Phase I, screening studies. Final report. SRI International, Menlo Park, CA.

Spanggord RJ, Mabey WR, Chou TW, Lee S, Alferness PL, Tee DS, Mill T (1983) Environmental fate studies of HMX. Phase II, detailed studies. Final report. SRI International, Menlo Park, CA.

Spiker JK, Crawford DL, Crawford RL (1992) Influence of 2,4,6-trinitrotoluene (TNT) concentration on the degradation of TNT in explosive-contaminated soils by the white rot fungus *Phanerochaete chrysosporium*. Appl Environ Microbiol 58:3199–3202.

SRC (1995a) Syracuse Research Corporation. (K_{oc} values cited in HSDB 1995b.)

SRC (1995b) Syracuse Research Corporation. Data calculated by SRC, contained in the Hazardous Substances Data Bank (HSDB), MEDLARS Online Information Retrieval System, National Library of Medicine. Retrieved 5/24/95.

Stephan CE, Mount DI, Hansen DJ, Gentile JH, Chapman GA, Brungs WA (1985) Guidelines for deriving national water quality criteria for the protection of aquatic organisms and their uses. PB85-227049. U.S. Environmental Protection Agency, Washington, DC.

Stidham BR (1979) Analysis of wastewater for organic compounds unique to RDX/HMX manufacturing and processing. Final report. AD AO85765. Holston Defense Corporation, Kingsport, TN.

Stillwell JM, Fischer MA, Margard WL, Matthews MC, Sherwood BE, Stanford TB (1977) Toxicological investigations of pilot treatment plant wastewaters at Holston Army Ammunition Plant. Final report, AD A042601. Battelle Columbus Laboratories, Columbus, OH.

Sullivan JH Jr, Putnam HD, Keirn MA, Pruitt BC, Swift DR, McClave JT (1978) Winter field surveys at Volunteer Army Ammunition Plant, Chattanooga, Tennessee. Final report. ADA 055 901. Water and Air Research, Inc., Gainesville, FL.

Sullivan JH Jr, Putnam HD, Keirn MA, Pruitt BC Jr, Nichols JC, McClave JT (1979) A summary and evaluation of aquatic environmental data in relation to establishing water quality criteria for munitions unique compounds. AD AO87683. U.S. Army Medical Research and Development Command, Ft. Detrick, MD.

Sunahara GI, Renoux AY, Dodard S, Paquet L, Hawari J, Ampleman G, Lavigne J, Thiboutot S (1995) Optimization of extraction procedures for ecotoxicity analyses: use of TNT-contaminated soil as a model (abstract). National Research Council of Canada, Montreal, Quebec, Canada.

Suter GW, Tsao CL (1996) Toxicological benchmarks for screening potential contaminants of concern for effects on aquatic biota: 1996 revision. ES/ER/TM-96/R2. Oak Ridge National Laboratory, Oak Ridge, TN.

Tabek HH, Chambers CW, Kabler PW (1964) Microbial metabolism of aromatic compounds. I. Decomposition of phenolic compounds and aromatic hydrocarbons by phenol-adapted bacteria. J Bacteriol 87:910–919.

Thompson PL, Ramer LA, Guffey AP, Schnoor JL (1997) Decreased transpiration in poplar trees exposed to 2,4,6-trinitrotoluene. Environ Toxicol Chem 17:902–906.

Todd Q, Finger F, Turner R, Morley D (1989) Delivery Order 8, Louisiana Army Ammunition Plant: updated remedial investigation. Roy F. Weston, West Chester, PA. U.S. Army Toxic and Hazardous Materials Agency, Aberdeen Proving Ground, MD.

Toussaint MW, Shedd TR, van der Schalie WH, Leather GR (1995) A comparison of standard acute toxicity tests with rapid-screening toxicity tests. Environ Toxicol Chem 14:907–915.

Triegel EK, Kolmer JR, Ounanian DW (1983) Solidification and thermal degradation of TNT waste sludges using asphalt encapsulation. In: National Conference on Management of Uncontrolled Hazard Waste Sites. Hazardous Materials Controls Research Institute, Silver Spring, MD, pp 270–274. (Cited in ATSDR 1995a.)

Tucker WA, Dose EV, Gensheimer GJ (1985) Evaluation of critical parameters affecting contaminant migration through soils. Final report. U.S. Army Toxic and Hazardous Materials Agency, Aberdeen Proving Ground, MD.

U.S. Army (1967) Industrial Medical and hygiene considerations: trinitrotoluene (TNT). Regulation No. 40-3, Department of the Army Material Development and Readiness, Command, Alexandria, VA.

U.S. Army (1986) Demilitarization of conventional ordnance: priorities for data-base assessments of environmental contaminants. U.S. Army Medical Research and Development Command, Fort Detrick, MD.

U.S. Army (1987a) Conventional weapons demilitarization: a health and environmental effects database assessment. AD A2205888. U.S. Army Medical Research and Development Command, Fort Detrick, MD.

U.S. Army (1987b) Louisiana Army Ammunition Plant remedial investigation report. AMXTH-IR-CR-87100. U.S. Army Toxic & Hazardous Materials Agency, Aberdeen Proving Ground, MD.

U.S. Army (1989) Delivery Order No. 8, Louisiana Army Ammunition Plant: updated remedial investigation. Performed by Roy F. Weston, Inc., West Chester, PA, for the U.S. Army Toxic & Hazardous Materials Agency.

U.S. Army (1990) Phase I results report: remedial investigation, manufacturing area, Joliet Army Ammunition Plant, Illinois. Vol. 1. AD B148444. U.S. Army Toxic and Hazardous Materials Agency, Aberdeen Proving Ground, MD.

USACHPPM (1994) U.S. Army Center for Health Promotion and Preventive Medicine. Field survey No. 75-23-YS50-94. Health risk assessment of consuming deer from Aberdeen Proving Ground, Maryland (draft). Toxicology Division, USACHPPM, Aberdeen Proving Ground, MD.

USEPA (1975) Methods for acute toxicity tests with fish, macroinvertebrates, and amphibians. 660/3-75-009. Ecological Research Series, Washington, DC.

USEPA (1979) Toxic substances control act premanufacture testing of new chemical substances. Fed Reg 44:16257–16259 (March 16, 1979).

USEPA (1985) Health and environmental effects profile for dinitrobenzenes. EPA/600/X-85/361, ECAO-CIN-P141. Prepared by the Office of Health and Environmental Assessment, Environmental Criteria and Assessment Office, Cincinnati, OH.

USEPA (1986) Quality criteria for water, 1986. EPA 440/5-86-001, PB87-226759. National Technical Information Service, Springfield, VA.

USEPA (1989) Health and environmental effects document for RDX cyclonite. ECAO-CIN-GO78. Office of Solid Waste and Emergency Response, Washington, DC.

USEPA (1993a) Proposed water quality guidance for the Great Lakes System. Fed Reg 58:(72)20802–21047, April 16, 1993. (See also EPA/822/R-93/006, Office of Science and Technology, Washington, DC.)

USEPA (1993b) Technical basis for deriving sediment quality criteria for nonionic organic contaminants for the protection of benthic organisms by using equilibrium partitioning. EPA-822-R-93-011. Office of Water, Washington, DC.

USEPA (1993c) Integrated risk information system (IRIS) Online. 2,4,6-Trinitrotoluene. Office of Health and Environmental Assessment, Cincinnati, OH.

USEPA (1993d) Integrated Risk Information System (IRIS) Online. *m*-Dinitrobenzene. Office of Health and Environmental Assessment, Cincinnati, OH.

USEPA (1993e) Integrated Risk Information Systems (IRIS) Online. Octahydro-1,3,5,7-tetranitro-1,3,5,7-tetrazocine. Office of Health and Environmental Assessment, Cincinnati, OH.

USEPA (1993f) Integrated Risk Information System (IRIS) Online. RDX. Office of Health and Environmental Assessment, Cincinnati, OH.

USEPA (1995a) Great Lakes water quality initiative technical support document for wildlife criteria. EPA-820-B-95-009. Office of Water, Washington, DC.

USEPA (1995b) Health effects assessment summary table. Annual FY-1995. EPA 540/R-95-036. Prepared by the Office of Health and Environmental Assessment, Environmental Criteria and Assessment Office, Cincinnati, OH, for the Office of Emergency and Remedial Response, Washington, DC.

USEPA (1996) Ecotox thresholds. EPA 540/F-95/038, PB95-96324. Intermittent Bull 3: 1–12. Office of Solid Waste and Emergency Response, Washington, DC.

USEPA (1997) Integrated Risk Information System (IRIS) Online. 1,3,5-Trinitrobenzene. Office of Health and Environmental Assessment, Cincinnati, OH.

van der Schalie W (1983) The acute and chronic toxicity of 3,5-dinitroaniline, 1,3-dinitrobenzene, and 1,3,5-trinitrobenzene to freshwater aquatic organisms. AD A138408. U.S. Army Medical Bioengineering Research and Development Laboratory, Fort Detrick, MD.

Veith GD, Macek KJ, Petrocelli SR, Carroll J (1979) An evaluation of using partition coefficients and water solubility to estimate bioconcentration factors for organic chemicals in fish. Fed Reg 44:15973.

Veith GD, Macek KJ, Petrocelli SR, Carroll J (1980) An evaluation of using partition coefficients and water solubility to estimate bioconcentration factors for organic chemicals in fish. In: Eaton JG, Parrish P, Hendricks AC (eds) Aquatic Toxicology. ASTM STP 707. American Society for Testing and Materials, Philadelphia, PA, pp 116–129.

von Oepen B, Kordel W, Klein W (1991) Sorption of nonpolar and polar compounds to soils: processes, measurements and experience with the applicability of the modified OECD-guideline 106. Chemosphere 22:285–304.

Walsh ME (1990) Environmental transformation products of nitroaromatics and nitramines: literature review and recommendations for analytical development. ADA 220 610. U.S. Army Cold Regions Research and Engineering Laboratory, Hanover, NH.

Walsh ME Jenkins TF (1992) Identification of TNT transformation products in soil. ADA 225 308. U.S. Army Corps of Engineers, Cold Regions Research & Engineering Laboratory, Hanover, NH.

Weitzel RL, Simon BP, Jerger DE, Schenk JE (1975) Aquatic field study at Iowa Army Ammunition Plant. Final report. AD AO14 300. Environmental Control Technology Corporation, Ann Arbor, MI.

Wentzel RS, Hyde RG, Jones WE (1979) Problem definition study on 1,3-dinitrobenzene, 1,3,5-trinitrobenzene and di-*n*-propyl adipate. AD A099732. U.S. Army Medical Research and Development Command, Fort Detrick, MD.

Will ME, Suter GW II (1995a) Toxicological benchmarks for screening potential contaminants of concern for effects on terrestrial plants: 1995 revision. ES/ER/TM-85/R2. Oak Ridge National Laboratory, Oak Ridge, TN.

Will ME, Suter GW II (1995b) Toxicological benchmarks for screening potential contaminants of concern for soil and litter invertebrates and heterotrophic processes. ES/ER/TM-126/R1. Oak Ridge National Laboratory, Oak Ridge, TN.

Won WD, DiSalvo LH, Ng J (1976) Toxicity and mutagenicity of 2,4,6-trinitrotoluene and its microbial metabolites. Appl Environ Microbiol 31:576–580.

Won WD, Heckly RJ, Glover DJ, Hoffsommer JC (1974) Metabolic disposition of 2,4,6-trinitrotoluene. Appl Microbiol 27:513–516.

Yasuda SK (1970) Separation and identification of tetryl and related compounds by two-dimensional thin-layer chromatography. J Chromatogr 50:453–457.

Yinon J (1990) Toxicity and Metabolism of Explosives. CRC Press, Boca Raton, FL.

Yinon J, Hwang DG (1987) Applications of liquid chromatography-mass spectrometry in metabolic studies of explosives. J Chromatogr 394:253–257.

Zambrano A, Mandovano S (1956) Urinary excretion of picric acid, picramic acid, and of sulfoconjugation products in experimental tetryl poisoning. Folia Med 39:162–171.

Manuscript received January 31, 1997; accepted June 19, 1998.

Appendix I: List of Abbreviations and Acronyms

2-ADNT	2-Amino-4,6-dinitrotoluene
AAP	Army Ammunition Plants
ACR	Acute/chronic ratio
BCF	Bioconcentration factor
bw	Body weight
CCC	Criterion Continuous Concentration
CMC	Criterion Maximum Concentration
Cf	Dietary screening benchmark
CV	Chronic Value (geometric mean of the NOEC and LOEC)
Cw	Drinking water screening benchmark
DNA	3,5-Dinitroaniline
DNB	1,3-Dinitrobenzene
EC_{50}	Effective concentration for 50% of test organisms
f	Food factor (the amount of food consumed/unit bw/d)
FAV	Final Acute Value
FACR	Final Acute/Chronic Ratio
FCV	Final Chronic Value
FCM	Food chain multiplying factor
f_{oc}	Fraction organic carbon in soil or sediment
FPV	Final Plant Value
GMAV	Genus Mean Acute Value
HMX	Octahydro-1,3,5,7-tetranitro-1,3,5,7-tetrazocine
K_d, K_p	Soil or sediment adsorption or partition coefficient
K_{oc}	Soil or sediment adsorption coefficient based on organic carbon

K_{ow}	Octanol/water partition coefficient (also called P)
LAP	Load, assemble, and pack plants
LC_{50}	Lethal concentration for 50% of test organisms (median lethal concentration)
LOAEL	Lowest-observed-adverse-effect level
LOEC	Lowest-observed-effect concentration
NOAEL	No-observed-adverse-effect level
NOEC	No-observed-effect concentration
RDX	Hexahydro-1,3,5-trinitro-1,3,5-triazine
RFD	Reference dose
SACR	Secondary acute/chronic ratio
SAV	Secondary Acute Value
SMAV	Species Mean Acute Value
SMC	Secondary Maximum Concentration
SCV	Secondary Chronic Value
SCC	Secondary Continuous Concentration
SQB	Sediment Quality Benchmark
SQB_{oc}	Sediment Quality Benchmark normalized to sediment organic carbon
SQC	Sediment Quality Criteria
tetryl	*N*-Methyl-*N*,2,4,6-tetranitroaniline
TNB	1,3,5-Trinitrobenzene
TNT	2,4,6-Trinitrotoluene
ω	Water factor (the amount of water consumed/unit bw/d)
WQC	Water Quality Criterion

Appendix II: Army Ammunition Plants (AAP) and Other Military Sites

Aberdeen Proving Ground, Edgewood, Maryland
Alabama AAP, Talladega County, Alabama
Camp Shelby, Hattiesburg, Mississippi
Chickasaw Ordnance Works, Chickasaw, Tennessee
Cornhusker AAP, Grand Island, Nebraska
Eagle River Flats, Fort Richardson, Anchorage County, Alaska
Fort Wingate Army Depot, Gallup, New Mexico
Hastings East Park, Hastings, Nebraska
Hawthorne AAP, Mineral County, Nevada
Hawthorne Naval Ammunition Depot, Hawthorne, Nevada
Holston AAP, Kingsport, Tennessee
Iowa AAP, Middletown, Iowa
Joliet AAP, Joliet, Illinois
Kansas AAP, Parsons, Kansas
Lexington-Bluegrass Depot, Lexington, Kentucky
Lone Star AAP, Texarkana, Texas
Louisiana AAP, Shreveport, Louisiana

Milan AAP, Milan, Tennessee
Naval Facility, Kitsap County, Washington
Naval Surface Warfare Center, Crane, Indiana
Nebraska Ordnance Works, Mead, Nebraska
Newport AAP, Newport, Indiana
Picatinny Arsenal, Rockaway Township, New Jersey
Radford AAP, Radford, Virginia
Raritan Arsenal, Raritan, New Jersey
Ravenna AAP, Ravenna, Ohio
Sangamon Ordnance Plant, Sangamon County, Illinois
Savanna Army Depot, Savanna, Illinois
Umatilla Army Depot, Umatilla, Oregon
VIGO Chemical Plant, Terre Haute, Indiana
Volunteer AAP, Chattanooga, Tennessee
Weldon Springs Ordnance Works, St. Charles County, Missouri
West Virginia Ordnance Works, Point Pleasant, West Virginia

Rev Environ Contam Toxicol 161:157–200

Aquatic Biotoxins: Design and Implementation of Seafood Safety Monitoring Programs

Douglas L. Park, Sonia E. Guzman-Perez, and Rebeca Lopez-Garcia

Contents

I. Introduction

When an issue on food safety is considered, it is important to state that absolute safety is not possible. Scientists and consumers alike concede that risks are associated with food as the result of compounds of chemical or microbiological

Communicated by George W. Ware

D.L. Park (✉)·S.E. Guzman-Perez·R. Lopez-Garcia
Louisiana State University, Department of Food Science, 111 Food Science Building, Baton Rouge, LA 70803, U.S.A.

Editor's Note: In this manuscript, we have deviated from the traditional referencing system to one using numbered references. This was done to enable simplified referencing within the massive data sets found in the tables. All references are both numbered and alphabetized in the References.

origin. In fact, foods considered safe under normal conditions would not qualify for a "seal of approval" guaranteeing 100% safety if they were consumed in excessive quantities or used in an unusual manner. Relative food safety can be defined as the practical certainty that injury or damage will not result from the ingestion of a food or ingredient used in a reasonable and customary manner and quantity (57, 58).

The economic consequences of unsafe food have not been systematically quantified; however, economic losses induced by contaminated food are substantial. The losses involve the cost/value of food spoiled or destroyed as a result of the contamination, the costs of treating resultant diseases, and the losses induced by morbidity, disability, or mortality. Additionally, because of food hazards identified, food intended for export can be rejected by importing countries, thus resulting in national exchequer losses for exporting nations (21). Food is an essential element in human life, but food should not only be available in sufficient quantity, it should also be nutritious, safe, and wholesome (21). Fishery products are generally attractive, healthy, and nutritive. Unfortunately, some obstacles are in opposition to the best use of marine resources, either industrial contamination or naturally occurring toxins (from bacteria, algae, or seafood decomposition) (82).

Filter-feeding bivalve shellfish use microscopic planktonic algae as a source of food. When planktonic algae proliferate they form algal blooms. However, these algal blooms may become harmful, affecting the economy of surrounding areas and causing human health impacts (62). About 5000 species of marine phytoplankton are estimated to exist; however, only about 300 can discolor the surface of the sea and only about 40 have been identified to produce potent toxins that can enter the food chain through the consumption of fish and shellfish to humans (151). The term "red tide" is used when the algae grow in such abundance that they change the color of the seawater to red, brown, or green; however, the term is misleading because not all water discolorations are toxic. Therefore, the proper term is harmful algal blooms (HABs) (5, 62). Although the organisms are often referred to as harmful algae, they also include cyanobacteria.

HABs are entirely natural phenomena, which have occurred for years. However, the past two decades have been marked with apparent increased frequency, intensity, and geographic distribution. This apparent increase in HABs can be the result of increased scientific awareness of toxic species, increased use of coastal waters for aquaculture, stimulation of plankton blooms by cultural eutrophication or unusual climatological conditions, and transport of the resting cysts of dinoflagellates either in ship ballast water or associated with translocation of shellfish stocks from one region to another (3, 5, 62).

Marine toxins occur most significantly in shellfish and finfish. Paralytic shellfish poisoning (PSP), diarrheic shellfish poisoning (DSP), amnesic shellfish poisoning (ASP), and neurotoxic shellfish poisoning (NSP) fit in the first category. Ciguatera fish poisoning (CFP) and pufferfish poisoning are associated with marine fish toxins. Although circumstances leading to human exposure to cyanobacterial (blue-green algal) toxins through drinking water do not follow the etiology of seafood poisoning listed, these toxins can be a serious public health concern.

This chapter provides information and guidelines on outlining criteria for the establishment of effective seafood safety monitoring programs for finfish and shellfish. Summary background information on occurrence of these poisonings, the source and chemical characteristics of responsible toxins, and analytical capabilities are provided to assist in outlining recommendations for appropriate monitoring programs.

II. Background

A. Occurrence

The intermediate accumulators of toxins, known as transvectors (140), can be divided into primary and secondary types. Primary transvectors accumulate toxins by ingesting the causative organism directly (filter-feeding shellfish, herbivorous fish, and detrital feeders) (123). Secondary transvectors are those species that consume primary transvectors; these are primarily higher carnivores (123).

Bivalves are the principal vectors of aquatic intoxication for humans. Shellfish accumulate the toxins via filter-feeding. When the algal bloom subsides, shellfish usually depurate the toxins naturally (82). The interactions between marine biotoxins and bivalve shellfish are complex and dynamic, varying between species and even within subpopulations. The rates of intoxication and detoxification of filter-feeding shellfish by toxic algae are species specific and are, in most instances, directly related to the toxic algal abundance and availability to the animals (41, 148). The rate of detoxification is highly dependent on the site of toxin storage within the animal; i.e., toxins in the gastrointestinal tract (e.g., *Mytilus*) are eliminated much more readily than toxins bound to tissues (e.g., *Placopecten, Spisula*, and *Saxidomus*) (148).

Table 1 summarizes the occurrence of the toxins and the worldwide location of the poisonings. The map in Fig. 1 summarizes the location of the most common seafood-related diseases from aquatic biotoxins.

B. Source of the Toxins

Not all HABs are associated with toxic species and the contamination of shellfish; they can also result from concentrations of nontoxic dinoflagellates or ciliates (172). Toxic algal blooms present not only a public health hazard but a major economic threat as well. Blooms may affect the fisheries and culture efforts by rendering products toxic and thus unmarketable, either by directly killing the shellfish (by oxygen deficiency) or by what is known as banning (74, 147). Unusual bloom behavior appears to occur in two forms, either in a periodic short burst or as a persistent although erratic growth in microbial populations (13).

Marine dinoflagellates are the most prolific biotoxin producers known, and therefore are of primary concern. However, other divisions of algae, bacteria, and cyan-

Table 1. Worldwide location and vectors of aquatic biotoxins.

Disease	Place	Vectors
Paralytic shellfish poisoning	Temperate regions (4, 12): • East and West Coasts of Canada • East and West Coasts of U.S.A. • Europe Tropical regions (47, 48, 109): • Central America • South America • Southeast Asia	• Crabs • Oysters • Scallops • Mussels • Marine snails • Top and turban shells (107)
Diarrheic shellfish poisoning	• Japan • Europe • North America (56, 128)	• Clams • Mussels • Oysters • Scallops (41)
Neurotoxic shellfish poisoning	• Florida coastline in the Caribbean (8, 9)	• Clams • Other bivalves (82)
Amnesic shellfish poisoning	• Prince Edward Island • Southern California to Alaska (159, 171)	• Anchovies • Crabs • Mussels • Razor clams (159, 171)
Ciguatera fish poisoning	Tropical regions: • Caribbean • Indian Ocean • Pacific rim Temperate regions: • Subtropical North Atlantic (46, 54)	• Amberjack • Barracuda • Grouper • Moray eel • Parrot fish • Snapper • Spanish mackerel (63)
Pufferfish poisoning	Primarily: • Japan Other regions: • China Ocean • Indian Ocean • Mediterranean Sea • Pacific Ocean (105)	• Atelopoid frogs • Octopuses • Pufferfish • Shellfish • Starfish • Taricha salamanders (97)
Cyanobacterial poisoning	• 27 States in U.S.A. • 16 Countries in Europe (25)	• Water (26)

Fig. 1. Global distribution of common seafood-related diseases from aquatic biotoxins. *A*, amnesic shellfish poisoning; *C*, ciguatera fish poisoning; *D*, diarrheic shellfish poisoning; *N*, neurotoxic shellfish poisoning; *P*, paralytic shellfish poisoning.

obacteria also are documented to produce biotoxins that impact human health (43).

Mostly because of their mobility, dinoflagellates were classified as a phylum of protozoans, but they are more properly classified as algae. It has been proposed that dinoflagellates are evolutionary bonds between prokaryotes (e.g., blue-green algae) and eukaryotes (e.g., green or brown algae) (146). Most of the definitive cell biology for toxic dinoflagellates has yet to be observed, understood, and described (181). Dinoflagellates are eukaryotic marine algae belonging to the division Phyrrophyta (152). They are among the principal producers of organic matter and are a major element of the marine phytoplankton.

The explosive growth of dinoflagellates is believed to be caused by a combination of optimum temperature, nutrient concentration, salinity, sunlight, freshwater runoff, and water stability. The cells of toxic dinoflagellates seem to prefer a low light environment, just at the interface of the low light–high nutrient zone. Current knowledge indicates that although chemical and biotic factors are important for *in situ* growth of dinoflagellate cells, convergence by thermal and tidal fronts is essential for cell accumulation and bloom development. Endemic dinoflagellate species can be accidentally introduced when their cysts are released with the ballast tank water and sediments of bulk cargo vessels (60).

Another way of distributing algae is with the transfer of shellfish stocks from one region to another (61).

The regions of oceans most severely affected by HABs have been waters that are relatively pristine (181). Areas such as Hong Kong harbor, the Seto Inland Sea in Japan, and the Northern European coastal waters present evidence that "cultural eutrophication" from domestic, industrial, and agricultural wastes are able to stimulate dangerous algal blooms (61). Furthermore, researchers have noticed correlation between elevated water contamination and increased dinoflagellate blooms (181). However, the apparent spreading of toxic blooms is still not clear (78).

Table 2 summarizes the principal organisms implicated with the accumulation of aquatic biotoxins in seafoods and drinking water.

C. Toxins

Marine toxins have drawn scientists' attention because of their involvement in human intoxication and the socioeconomic impacts brought by those incidents. It is imperative to elucidate the chemical structures of the toxins, not only for understanding the molecular basis of mechanisms of action but also for designing proper countermeasures such as detection, determination, and therapeutic methods (179). Most of the known biotoxins (Table 3) are heat stable, odorless, colorless, and otherwise completely undetectable by the human senses (43).

Although freshwater cyanobacterial toxins are not considered to be seafood toxins, there are several points of similarity making it useful to consider them along with these other toxins: (1) both are water-based biotoxins, (2) both are produced by microorganisms, (3) while retained within the cells to varying degrees both groups are exotoxins, (4) the toxins are fast-acting neuro- or organo-compounds absorbed via the oral route, and (5) certain cyanobacteria toxins have structural/functional similarities to certain paralytic shellfish toxins, especially saxitoxin and neosaxitoxin (23).

D. Analytical Methodology

Marine toxins pose a significant challenge to the analytical chemist. They range from the very polar saxitoxin-based compounds responsible for PSP to lipophilic toxins such as ciguatoxin and okadaic acid. The potent toxins demand that analytical methods provide high sensitivity, and complex shellfish and finfish tissue matrices require high selectivity of separation and detection (127).

Structure variation of the diverse toxins poses a serious problem for the development of chemical assays and methods of confirmation of identity. Mass spectrometry holds great promise for low-level toxin detection (121). The capabilities of immunological technology for assessment of toxins in seafood products are promising and show the highest potential for wide use for seafood monitoring programs (70). Immunochemical procedures allow for screening a larger number of samples in a shorter time than that needed for running chemical procedures such as HPLC (51).

Table 2. Organisms implicated with aquatic biotoxins.

Disease	Selected organisms implicated
Paralytic shellfish poisoning	Dinoflagellates (59): • *Alexandrium* spp. (formerly *Gonyaulax, Protogonyaulax*) • *Gymnodinium catenatum* • *Pyrodinium bahamense*
Diarrheic shellfish poisoning	Dinoflagellates (42, 84): • *Dinophysis fortii, D. rotundata, D. acuta* • *Prorocentrum lima, P. concavum*
Neurotoxic shellfish poisoning	Dinoflagellate (1, 9): • *Ptychodiscus brevis* (formerly *Gymnodinium breve*)
Amnesic shellfish poisoning	Diatoms (1, 52, 160, 176): • *Nitzchia pungens* f. *multiseries, N. pseudodelicatissima* • *Pseudonitzchia australis* Red algae: • *Chondria armata* and *Chondria bailayena*
Ciguatera fish poisoning	Dinoflagellates (72, 157): • *Gambierdiscus toxicus* (confirmed) • *Coolia monotis* • *Ostreopsis ovata, O. siamensis* • *Prorocentrum lima, P. concavum*
Pufferfish poisoning	Bacteria (45): • *Acinetobacter* spp. • *Actinomycetes* spp. • *Aeromonas* spp. • *Alteromonas* spp. • *Bacillus* spp. • *Flavobacterium* spp. • *Micrococcus* spp. • *Moraxella* spp. • *Plesiomonas* spp. • *Vibrio* spp.
Cyanobacterial poisoning	Blue-green algae (25, 26): • *Anabaena* spp. • *Aphanizomenon* spp. • *Microcystis* spp. • *Nodularia* spp. • *Oscillatoria* spp.

Table 3. Principal toxins associated with seafood-related diseases.

Disease	Toxins
Paralytic shellfish poisoning	Saxitoxin Neosaxitoxin Gonyautoxins 1 to 4 Decarbamoyl saxitoxin, decarbamoyl neosaxitoxin, and decarbamoyl gonyautoxins 5 to 6 C1 to 4 (64, 179)
Diarrheic shellfish poisoning	Okadaic acid Dinophysistoxins 1, 2, 3 Pectenotoxins 1, 2, 3, 4, 6 Yessotoxins (40, 179)
Neurotoxic shellfish poisoning	Brevetoxins type A: Brevetoxin-1, -7, -10 Brevetoxins type B: Brevetoxin-2, -3, -5, -6, -8, -9 (8, 9)
Amnesic shellfish poisoning	Domoic acid Isodomoic acids A to F Isodomoic acid C5′ diastereoisomer Domoilactones A and B (174, 175, 176)
Ciguatera fish poisoning	Ciguatoxins (confirmed) Gambiertoxins (confirmed) Maitotoxins Scaritoxin Okadaic acid (85, 87, 99, 141)
Pufferfish poisoning	Tetrodotoxin (97)
Cyanobacterial poisoning	Hepatotoxins: Microcystins Nodularin Neurotoxins: Anatoxin-a Anatoxin-a(s) Homoanatoxin Saxitoxin Neosaxitoxin (16, 17, 25, 138, 149)

Table 4 summarizes current analytical techniques to determine aquatic biotoxins; Table 5 summarizes selected information on diseases associated with aquatic biotoxins.

III. International Programs

A number of international programs have been or are under development to study and manage HABs and their linkages to environmental changes in a manner consistent with the global nature of the phenomena involved (61).

A. Intergovernmental Oceanographic Commission (IOC) of UNESCO

The IOC, based on guidance from member nations, is encouraging the development of a comprehensive global ocean observing system to provide information needed for oceanic and atmospheric forecasting for ocean management by coastal nations. IOC also encourages research on global environmental change, related education, and training and technical assistance programs to ensure that all countries can participate and benefit from the effort.

The Harmful Algal Bloom Program (HAB) was initiated by IOC member states as an activity under the joint IOC-FAO Ocean Science in Relation to Living Resources Program (OSLR). Through a number of international workshops, a program plan was prepared to cover educational, scientific, and operational aspects of harmful algae. Program activities are developed continuously; and, since 1992, implemented, jointly by IOC member states, the IOC secretariat, and cosponsoring organizations through the IOC-FAO Intergovernmental Panel on Harmful Algal Blooms (IPHAB). A task team in aquatic biotoxins has been established to prepare documentation on chemical and toxicological characteristics of aquatic biotoxins.

The goal of IPHAB is to foster the effective management of and scientific research on HABs to understand their causes, predict their occurrences, and mitigate their effects. Educational activities include training courses on taxonomy, toxin determination, and monitoring of harmful algae. Individual study grants are offered, and the publication of a newsletter on toxic algae and algal blooms (Harmful Algae News), manuals, guides, and directories on various topics related to harmful marine microplankton are key components of the program. Scientific activities include an ICES-IOC working group on HAB dynamics, an SCOR-IOC working group on the physiological ecology of HABs, regional science planning workshops, and related pilot projects and workshops. Operational activities include initiatives directed toward improved resource protection, monitoring, public health, and seafood safety.

B. World Health Organization (WHO): International Program on Chemical Safety

The members of the International Program on Chemical Safety and the International Life Sciences Institute (76) have presented a list of priorities for risk

Table 4. Current analytical techniques available for the determination of selected aquatic biotoxin-related diseases.

Disease	Methodology	Comments
Paralytic shellfish poisoning	• Mouse bioassay (6, 170)	• Detection level = 40 μg/100 g. Lack of specificity and precision. Salts and metals interfere.
	• TLC-UV (20)	• Plates are sprayed with H_2O_2 to give fluorescent derivative. Difficult to quantify.
	• HPLC with postcolumn derivation to fluorescent compounds (14, 108)	• Detects saxitoxin, neosaxitoxin, gonyautoxins 1 to 6, decarbamoyl saxitoxin, and C1-C4 toxins. Detection level = 20–110 femtomoles.
	• HPLC with prechromatographic oxidation and mass spectrometric detection (7, 80, 81)	• Detects decarbamoyl saxitoxin, decarbamoyl neosaxitoxin and decarbamoyl gonyautoxins 2 and 3.
	• RIA (22)	
	• ELISA (31, 143)	• Detection level = 1 pg saxitoxin. Cross-reacts with neosaxitoxin, and gonyautoxins 2 and 3.
	• Receptor binding assay (33, 166)	• Detection level = 4 ng saxitoxin equivalents/mL.
	• Cytotoxicity assay (94, 95, 162)	• Detection level = 0.1 ng/10 μL or 2 μg/100 g tissue.
	• Capillary electrophoresis (156)	• Separation of underivatized toxins with detection by either a high sensitivity UV sensitivity UV detector or IOMS.
Diarrheic shellfish poisoning	• Mouse bioassay (170)	• Detects okadaic acid, dinophysistoxins 1 to 3. Detection level = 200 ng/g. 1 MU = 4 μg okadaic acid.
	• TLC (131, 170)	• Detection level = 1 μg.
	• Reverse-phase HPLC with UV detection (131)	• Detection level = 10 μg/mL.
	• HPLC with derivatization to fluorescent ester (ADAM reagent) (83, 84, 117, 131, 133)	• Detection level = 10 pg of okadaic acid derivative.
	• RIA (86)	

Table 4. Continued.

Disease	Methodology	Comments
	• HPLC-IOMS (119, 130)	• Detects okadaic acid, dinophysistoxin 1 and 3. Detection levels = 1 ng/g.
	• ELISA (28, 51, 144, 145, 165)	• DSP-Check detects okadaic acid, dynophysistoxin-2. Detection level = 20 ng/g.
	• S-PIA (51, 112, 114)	• Detects okadaic acid, dynophysistoxin-2. Detection level = 5 ng/g.
	• Cytotoxicity assay (37, 164)	• Detection level = 0.005 μg okadaic acid/mL.
	• Micellar electrokinetic chromatography with UV detection (18)	• Detection level = 40 pg okadaic acid.
	• HPLC-linked phosphatase radioassay (71)	• Based on inhibition of PP1 and PP2A. Detection level > 4 ng okadaic acid/mL extract.
	• Protein phosphatase assay (73)	• Detection level = 4 ng okadiac acid/mL.
Neurotoxic shellfish poisoning	• Mouse bioassay (9, 155)	• Is not highly sensitive nor specific.
	• Isocratic HPLC	
	• RIA (121, 122)	• Rapid, specific, and reproducible results, but not readily adapted for use in field. Detection level = 1 nM.
	• ELISA (9, 161)	• More rapid and sensitive than RIA, but cannot detect brevetoxin in fish (used to test dinoflagellate cells). Detection level = 0.4 pM.
	• Silica gel TLC (121)	• Excellent separation of all brevetoxins, but the sensitivity is a problem (>1 ppm.)
	• Reverse-phase liquid chromatography (82)	• Excellent separation of all brevetoxins. Detection level = 5 μg.
	• Cytotoxicity assay (94, 95)	• Simple, sensitive, can be alternative to animal testing. Detection level = 0.25 ng/10 μL.
	• Receptor binding assay (166)	• Continued effort still needed.

Table 4. Continued.

Disease	Methodology	Comments
Amnesic shellfish poisoning	• Mouse bioassay (158, 173)	• Not specific nor sensitive. Detection level > 40 μg domoic acid.
	• TLC (38, 131)	• Normal amino acids present in crude extracts should be removed. Detection level = 0.5 μg.
	• Reverse-phase HPLC with UV detection (125)	• Detection level = 10–80 ng/mL.
	• Direct HPLC with UV detection (79, 132)	• Detection level = 20–30 ng/g.
	• Reverse-phase gradient HPLC with fluorometric detection (120)	• Detection level = 15 pg/mL.
	• HPLC-IOMS (126)	• Detection level = 10–80 ng/mL.
	• GC-MS (118)	• Detection level = 1–500 μg/g wet tissue.
	• RIA (102)	• Still a research stage.
	• Receptor binding assay (166)	• Detection level = 0.001 μg/kg.
	• Capillary electrophoresis with UV detection (103, 129)	• Detection level similar to HPLC methods. Simple and rapid.
Ciguatera fish poisoning	• Mouse bioassay (36, 66, 93, 177)	• Detection level > 0.5 ppb ciguatoxin-1.
	• RIA (67)	• Not adapted for use in field.
	• ELISA (68)	• Lack of standards
	• S-PIA (112–115)	• High potential for screening.
	• HPLC with UV detection (93)	• Cannot be used with crude lipid extracts from fish. Detection level = 5 ng pure ciguatoxin-1.
	• HPLC-IOMS (92)	• Encouraging results, but needs more effort.
	• HPLC with fluorescence detection (44, 93, 180)	• Encouraging results.
	• NMR (87, 98)	

Table 4. Continued.

Disease	Methodology	Comments
	• Receptor-binding assay (166)	• High sensitivity, rapid, simple. Continued effort is still needed. Detection level = 1 pg ciguatoxin-1.
	• Cytotoxicity assay (94, 95)	• Highly sensitive and simple. Can be an alternative to animal testing; 10^4 fold more sensitive than mouse bioassay.
Pufferfish poisoning	• Mouse bioassay (106)	• 1 MU = 220 ng tetrodotoxin.
	• HPLC-fluorometry (106)	• Good linearity with mouse bioassay. More sensitive and specific. Detection level = 5 ng.
	• Mass spectrometry (106)	
	• TLC (106)	• Detection by spraying Weber reagent under UV light.
	• Electrophoresis (106)	
	• Capillary isotachophoresis (106)	
	• Immunoassay	• High specificity and sensitivity, further testing.
Cyanobacterial poisoning	• Mouse bioassay (136)	• Cannot detect low amounts of toxins and cannot distinguish between different types of neuro- or hepatotoxins.
	• HPLC (53)	
	• GC-electron-capture device (154)	
	• TLC (101)	
	• Fast atom bombardment MS (65)	• Expensive. Detection level ≈ 1 μg.
	• ELISA (19, 34, 35)	• The antibody reacts with all microcystins and nodularin.

ELISA, enzyme-linked immunosorbent assay; GC, gas chromatography; HPLC, high performance liquid chromatography; IOMS, ion spray mass spectrometry; MS, mass spectrometry; MU, mouse unit; NMR, nuclear magnetic resonance; RIA, radioimmunoassay; S-PIA, solid-phase immunobead assay; TLC, thin-layer chromatography; UV, ultraviolet.

Table 5. Summary of selected information on diseases associated with aquatic biotoxins.

	PSP	DSP	NSP
Vectors (Table 1)	Clams, mussels; toxin in digestive gland, siphon	Clams, mussels; toxin in digestive gland	Clams and other bivalves
Common geographic locations (Fig. 1)	North and South America, Europe and Asia	Heaviest around Japan and Europe. Some cases in U.S.A. and Canada	Mostly coast of Florida, Gulf of Mexico, New Zealand
Source of toxin (major species) (Table 2)	*Alexandrium* spp. *Gymnodinium catenatum* *Pyrodinium bahamense*	*Dinophysis* spp. *Prorocentrum lima* *Prorocentrum concavum*	*Ptychodiscus brevis*
Climatic zones	Tropical and temperate	Temperate	Tropical
Major toxins (Table 3)	Saxitoxin, neosaxitoxin, gonyautoxins, decarbamoyl toxins	Okadaic acid, dinophysistoxins, pectenotoxins, yessotoxins	Brevetoxins
Solubility	Water (124)	Fat (124)	Fat (124)
LD_{50} (ip. mouse, μg/kg)	saxitoxin: 9–11.6 (124)	Okadaic acid: 200; dinophysistoxin-1: 160; dinophysistoxin-3:500 pectenotoxin-1: 250; -2: 230; -3: 350; -4: 770; -6: 500 (178)	brevetoxin-1: 100 brevetoxin-2: 200 (8)
Mode of action	Block sodium channels (1)	Inhibit certain essential serine/threonine protein phosphatases (131)	Open sodium channels (8, 30)
Analytical procedures (Table 4)	Mouse bioassay, HPLC, ELISA, RIA, receptor binding assay	Mouse bioassay, HPLC, ELISA, S-PIA, cytotoxicity assay	Mouse bioassay, ELISA, RIA, HPLC, cytotoxicity and receptor binding assay

Table 5. Continued

ASP	CFP	Pufferfish poisoning	Cyanobacterial toxins
Mussels, clams	Most common in reef fish; toxin in gonads, viscera, liver, flesh	Most common in pufferfish; poison in liver, gonads, roe	Drinking and recreational water
Eastern Canada; Northeast and Northwest U.S.A	Pacific Ocean, Caribbean, Australia	Areas of Pacific around China and Japan, rare in U.S.A	Canada, U.S.A., Australia, Finland, United Kingdom, China, Japan, Brazil
Nitzcshia spp., *Pseudonitzcshia* spp., *Chondria armata*, *Chondria baileyana*	*Gambierdiscus toxicus*	Bacteria	*Aphanizomenon, Anabaena, Oscillatoria, Nodularia, Microcystis* spp.
Temperate	Tropical	Temperate	Tropical and temperate
Domoic acid, isodomoic acids, domoilactones	Ciguatoxins and gambiertoxins confirmed (89)	Tetrodotoxin	Microcystins, nodularin, anatoxin-a, saxitoxin, neosaxitoxin
Water (135, 176) domoic acid: 120 (55, 124)	Fat and water (124) Ciguatoxin-1: 0.25 Ciguatoxin-2: 2.3 Ciguatoxin-3: 0.9 (87)	Water (1) Tetrodotoxin: 7 (1)	Water (27, 124) Anatoxin-a: 200 Anatoxin-a(s): 20 Homoanatoxin: 50 Microcystin: 50–500 Nodularin: 30–50 (124)
Act as agonist to glutamate receptors (135, 176)	Open sodium channels (8)	Blocks sodium channels (1)	Neurotoxins prevent acetylcholinesterase from degrading acetylcholine. Hepatotoxins damage the liver by causing blood to pool in the liver. Can also inhibit protein phosphatases (27)
Mouse bioassay, HPLC, receptor binding assay	Mouse bioassay, RIA, ELISA, HPLC, S-PIA, binding assay, cytotoxicity assay	Mouse bioassay for PSP, HPLC, immunoassay, binding assay	Mouse bioassay, HPLC, ELISA, FABMS

Table 5. Continued

	PSP	DSP	NSP
Extent of problem	Local areas worldwide (1)	Potentially worldwide problem (1, 177)	Massive fish kills, environmental problems (1)
Compound stability	Heat stable; more stable at pH 7 and below (1)	Heat stable (1)	Heat stable (1)
Symptomatology in humans	Numbness, respiratory paralysis after eating; death within 2–12 hr; prognosis good after 24 hr (1, 61)	Abdominal pain, nausea, vomiting, diarrhea; rarely fatal. Okadaic acid and dinophysistoxins are tumor promoters (170, 177)	Symptoms like ciguatera from eating bivalves; asthma-like symptoms from sea spray; no deaths reported (1)
Level of concern in 100 g of meat	1 mg (sickness); 2 mg or more (death) (1)	20–40 μg okadaic acid (1)	Actual dose not known (1)
Treatment	No antidote; artificial respiration, rest (1, 14)	No specific treatment (1)	No specific treatment (1)
Control measures	80 μg/100 g meat (400 MU); good management of shellfish beds results in minimum cases (1, 150)	5 MU/100 g; Monitor for dinoflagellates, close shellfish beds (1, 170)	80 μg (20 MU)/100 g shellfish; close shellfish harvest during blooms (1)

ASP, amnesic shellfish poisoning; CFP, ciguatera fish poisoning; DSP, diarrheic shellfish poisoning; ELISA, enzyme-linked immunosorbent assay; FABMS, fast atom bombardment mass spectrometry; HPLC, high performance liquid chromatography; MU, mouse unit; NSP, neurotoxic shellfish poisoning; PSP, paralytic shellfish poisoning; RIA, radioimmunoassay; S-PIA, solid-phase immunobead assay.

evaluation of different marine biotoxins. These organizations have stressed the need for risk evaluation on an international scale. There appears to be general worldwide agreement on the need for measurements to control shellfish toxins in seafood, and many countries have taken legal action to ensure that phycotoxin-contaminated shellfish do not reach the consumer in any part of the world (148).

Table 5. Continued.

ASP	CFP	Pufferfish poisoning	Cyanobacterial toxins
Canada (1987): 103 cases and 3 deaths (1)	Largest seafood problem; 50,000 cases per year worldwide (1,10,134)	Important seafood in Japan; 100 cases, 50 deaths per year; rare elsewhere (1, 105)	In addition to rare acute lethal poisonings, dermatitis from fresh water contact is frequent (24)
Heat stable (1)	Heat stable (1)	Heat stable; more stable between pH 4 and 9 (1)	Heat labile
Vomiting, cramps, diarrhea, short-term memory loss and disorientation (1, 159)	Gastrointestinal, neurological, and cardiovascular; rarely fatal (69)	Very similar to PSP (1)	Dermatitis, skin necrotis, blisters, liver with hemorrhagic necrodis; no confirmed human deaths (96)
2 mg (1)	35 ng	Same as for PSP (1)	Not applicable
Rest, symptomatic; no antidote (1)	No effective antidote; mannitol may be effective (15, 110)	Same as for PSP (105)	Limited and largely unavailable for the neurotoxins (136)
2 mg domoic acid/100 g; close shellfish beds when domoic acid is detected in shellfish (1, 168)	0.35 ng ciguatoxin-1/g fish; no harvesting where history of toxic fish is found (1,39)	Education on identification of toxic species and fish preparation procedures (1)	Monitor for blooms; find alternative water supply for animals until the bloom is eliminated (27)

C. Food and Agriculture Organization (FAO)

The FAO has been collaborating with a number of member nations to develop coherent national strategies for food control, including drafting up-to-date food laws and regulations, strengthening technical facilities such as laboratories, inspecting and administrating services, and developing national actions and programs for the improvement of food handling practices at all levels of the food chain (21). Some countries monitor only for one or two classes of toxins and a limited number of species, while others have monitoring programs with a long list of species monitored. Denmark is the most active in this respect (169). The acceptable levels for the major toxins of concern may differ significantly between countries; thus, it would be desirable if international organizations would evaluate the hazards caused by marine phycotoxins to provide a common basis for risk assessment (148). FAO provides direction and support for developing nations on food safety surveillance programs.

D. International Union for Pure and Applied Chemistry (IUPAC)

The IUPAC Commission on Food Chemistry is composed of working groups whose activities focus on mycotoxins, aquatic biotoxins, elemental analysis, natural toxins (plant), halogenated hydrocarbon environmental contaminants, animal drugs, and effects of food processing. The Aquatic Biotoxins Working Group is the unit within the Commission charged with the responsibility of coordinating all IUPAC activities associated with contamination of food and water with aquatic biotoxins, including phycotoxins. The Commission sponsors a series of symposia on mycotoxins and phycotoxins. The Aquatic Biotoxins Working Group of the Commission initiates and conducts research on aquatic biotoxins in food and water, initiates and conducts method performance studies of selected methods of analysis for aquatic biotoxins in food and water, addresses the issue of criteria of purity and stability of individual toxins, collaborates with other international organizations involved in dealing with the consequences of human exposure to aquatic biotoxins, and plans and conducts symposia.

E. South Pacific Commission

The South Pacific Commission was formed in 1947 by the founding members: United Kingdom, Australia, New Zealand, France, United States of America, and The Netherlands. Australia continues to be the dominant force, providing about 33% of the annual budget. Member countries now number 27, which includes the founders (except The Netherlands) and most of the island nations in the South Pacific. The Commission is the older of two regional organizations which act to assure the economic and social stability of the island nations by creating mechanisms for meaningful relations between governments. This is accomplished through direct aid, grants for programs, and publication of technical papers, handouts, information documents, information circulars, reports of meetings, statistical bulletins, and newsletters on various topics.

F. Global Monitoring Programs

A greater emphasis on global seafood safety and algal bloom monitoring is under way. Considering both health and economic aspects of unsafe food, governments are providing increased attention to the prevention and control of food contamination, and national, regional, and local programs for food control are being strengthened and implemented. International assistance is greatly needed to improve the quality and safety of food supplies moving at global levels (21). For the food control system to contribute to the achievement of its goals, there must be a basic law, supplemented by detailed regulations and administered by a capable food control organization. It is known that new toxic phytoplankton species can be carried from one country to another in the ballast waters of commercial vessels (100). The potential for international transfer of toxic marine dinoflagellates and other nuisance organisms in shipping ballast water has been highlighted by several research groups in recent years (137).

IV. Guidelines for Monitoring Programs

A. Introduction

Shellfish and finfish have become important items in the world's food supply. Rapid air transportation with good refrigeration has widened local seacoast markets to many inland cities and areas throughout the world. Commercial growing and harvesting of shellfish for market has become a growing industry (142). For this reason, caution should be exercised when seafoods are transferred from one geographic region to another (182). Seafood monitoring programs have been established for selected shellfish susceptible to aquatic biotoxin contamination. For ciguatera, it will be difficult to control the consumption of fish in importing countries unless there is a program for monitoring fish harvesting areas and fish in the marketplace for ciguatera potential (116). In tropical islands where fish may be ciguatoxic, there is an accepted risk in eating reef fish. However, for fish that are exported from these areas, monitoring programs should include confirmation of the location of the catch, particularly if reef areas are fished where ciguatoxic fish are known to exist (39). A key component to this type of monitoring would be rapid, accurate, low-cost test procedures (116). Affordable tests for routine analysis for all export fish and shellfish are crucial to effective seafood safety monitoring programs (114).

Several factors that should be considered in the development of a food safety program in a given country include the public health significance of the contaminant, the effects on the countries economy, legal infrastructure, analytical resources, producer/consumer education, and the effectiveness of communication systems (109).

Risk evaluation is considered to be highly desirable for marine biotoxins to form a basis for regulation, but this has been judged difficult because of lack of reliable information on exposure and hazard (76). Risk management decisions are usually made with inadequate knowledge of all foregoing factors. The management plan can take the form of maximum tolerated levels that are legally enforceable, educational programs leading to a better understanding of how and where the toxicant occurs, with prevention of the occurrence or removal of the toxicant as the ultimate goal (111). Since high-risk aquatic biotoxins have been identified, short-term seafood safety programs should focus on the elimination of acute toxicity syndromes through the development of reliable analytical methods and the setting of regulatory limits. A long-term goal should be concerned with chronic toxicity and the development of decontamination procedures that would reduce the relative risk involved in the consumption of seafood.

An effective seafood safety program must include the ability to monitor fish and shellfish harvesting areas, the establishment of regulatory limits or a level of concern for the toxin(s), the ability to screen for suspect fishery products in the marketplace, and the management of unsafe fishing/harvesting areas and products (114) (Fig. 2). These programs are designed to identify toxic or high-risk products and divert them into lower-risk uses and allow acceptable products to proceed in commercial channels.

For most phycotoxins, the establishment of adequate seafood safety programs has been hampered by the lack of adequate standards and analytical methods that could be used for monitoring the presence or absence of these toxins in fish or shellfish at various points in commercial channels (170). With the exception of PSP, regulations established for aquatic biotoxins are limited. An international inquiry was undertaken in 1990 within the framework of a project of the IUPAC to obtain up-to-date information about worldwide marine phycotoxin legislation. As of January 1991, 21 countries had proposed regulations for marine phycotoxins (148, 168, 169), compared with 66 countries for mycotoxins (167) (Fig. 3).

B. Guidelines for the Establishment of Shellfish and Finfish Monitoring Programs

Monitoring has been defined as periodic sampling, organized to detect changes in the composition of the toxin-producing plankton and contaminated seafoods

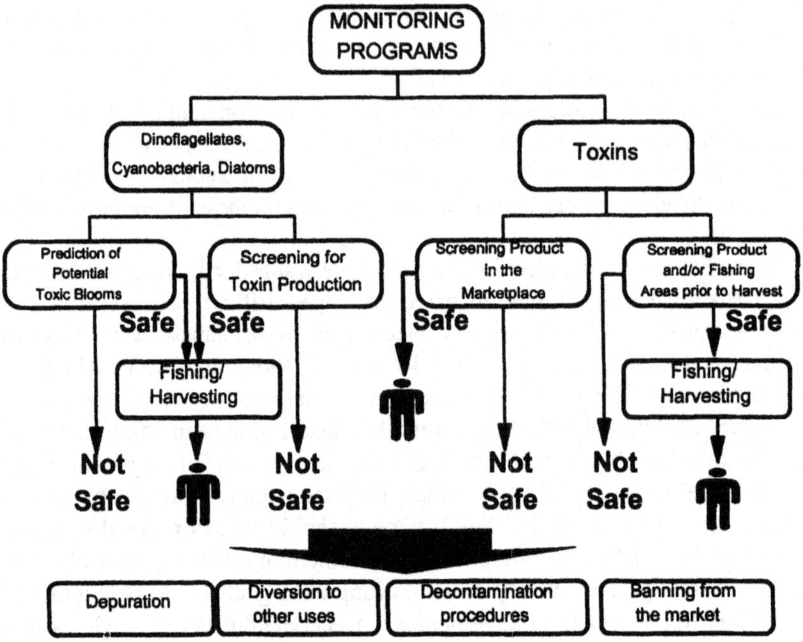

Fig. 2. Components and establishment of a seafood safety monitoring program.

Fig. 3. Geographic location of countries with marine phycotoxin legislation (*dark areas*) and of countries without marine phycotoxin legislation (*white areas*). Reprinted by permission of CNEVA (France). "Worldwide Regulations for Marine Phycotoxins," CNEVA—Actes du Colloque sur les Biotoxines Marines, by van Egmond, van den Top and Speijers, 1991, pg. 169.

(both preharvest and postharvest) and the geographic distribution thereof. Sampling represents one of the most important steps in any monitoring program. Information suggesting the accumulation of toxins in seafood must be followed up by intensive sampling (104).

Monitoring for algal blooms may be considered an important tool to avoid public health problems caused by the accumulation of marine toxin seafoods. Early detection would allow officials to warn people of the impending blooms and forewarn mariculturists or fishermen of pending economic disasters. Because most blooms originate offshore, satellite imagery or satellite-tracked monitoring buoys can assist in early detection of blooms (147).

Many of the components of seafood safety monitoring programs are similar for shellfish and finfish operations. The major difference, however, is that shellfish are not mobile whereas finfish can migrate between harvesting areas previously identified as either safe or potentially toxic. The components for the establishment of a monitoring program are summarized in Table 6.

Step 1: Establish the sampling plan.

The sampling plan is composed of several components including the frequency and location of sample collection, preparation of test portions for analysis, and relative errors associated with sampling along with the error inherent to the analytical method utilized. Special efforts must be undertaken to ensure that the sample represents the product lot or sampling location,

Table 6. Summary of components of the development of a safety monitoring program.

Step	Recommendation
1. Establish the sampling plan	Determine frequency and location of sample collection and its preparation for analysis.
2. Determine algal species, commodities, and toxin(s) of concern for the program	Determine frequency and occurrence of toxicoses as well as commodities involved and pattern of seafood consumption in the area.
3. Establish analytical capabilities applicable to the monitoring program selected	Evaluate laboratory equipment and trained personnel available in the area. Participate in specialized training programs. Validate methods and obtain standards and standard material where available.
4. Establish regulatory limits	Consider regulations in force in other countries as examples.
5. Establish regulatory policy for violative product	Address appropriate action for violative products
6. Distributive information	Disseminate information through the public media and in local areas.

including potential variation between individual specimens or toxin distribution within each fish or shellfish.

In the United States, the monitoring of marine biotoxins or associated phytoplankton in seafood and the environment has been primarily the responsibility of state-sponsored programs (4). Several are currently in place, such as in Maine, Massachusetts, Washington, and California. In a phytoplankton monitoring program in Nova Scotia, Canada, the growers are responsible for taking the water samples and delivering them to the official analytical laboratory. This procedure has greatly facilitated the transport of the samples (29). Examples of management programs such as those presented by the U.S. Food and Drug Administration (49, 50), Cembella and Todd (32), Trusewich et al. (163), and Carmichael and Falconer (26) can assist in the development of monitoring programs for fish, shellfish, and cyanobacteria.

Step 2: Determine algal species, commodities, and toxin(s) of concern for the program.

There is no general rule to define harmful concentration of cells in an algal bloom, the concentration being species specific. New species of harmful algae are detected continuously (2). It is important to identify the commodities of concern and the pattern of consumption within a specific country to determine the level of concern and develop an adequate program suitable for that specific area. A survey of seafood consumed, pattern of consumption, and production practices, i.e., harvesting, storage, and commercial pathways, is recommended.

- Shellfish
  For shellfish commodities, it is necessary to determine the mode of production, whether the product harvested directly or produced by aquaculture facilities. Seafood-related diseases are usually commodity- and geographic location specific, e.g., PSP and DSP in molluscs and TTX in pufferfish.
- Finfish
  In some countries, common ciguateric fish are normally avoided, but where alternative food sources are not available, they are readily consumed (39). Development of pelagic and deep-slope fisheries should be pursued, so that the resource base is widened and reef fishes are not the sole source of fish protein. The South Pacific Commission has been instrumental in successfully introducing the simple technology required to catch deep-slope fish (nontoxic) in a number of island nations (39).

Fish that are sent for export should be scrutinized and checked as to the location of the catch, particularly if reef areas with toxic fish are known to exist (39). In some areas, it might be appropriate to screen all potentially ciguateric reef fish, whereas in other areas only the high-risk species presently marketed may need to be screened (91).

Step 3: Establish analytical capabilities applicable to the monitoring program selected.

With respect to testing fishery products, the type of analytical methods used in a monitoring program will vary according to the purpose, i.e., a screening method versus a method of confirmation of identity. For a method to be of value in screening marketplace products, it must (1) be easy to use and interpret; (2) be rapid, i.e., able to test a large number of samples in a short period of time; (3) accurately differentiate between toxic and nontoxic samples at the level of concern; (4) be of low cost; and (5) where feasible, provide for a means of confirmation of identity (112). Depending on the monitoring program, it is often useful to utilize a combination of market screening procedures followed by a confirmation of identity of samples testing positive. To set up a monitoring program, it is necessary to understand how products become toxic and develop analytical techniques for each critical point (113).

The selection of the method used is influenced greatly by the product that is being analyzed and at what point in commercial channels it is tested, i.e., can the product be tested before it becomes a perishable product (before harvest as with shellfish, for example) or must it be tested in the postharvest commercial channels when the product has a finite shelf life. The time required for testing purposes can influence significantly which methods are used, i.e., bioassays or chemical methods requiring hours or days to complete versus immunochemical methods which often take less time (115). For monitoring products with a limited shelf life, lengthy testing procedures would be prohibitive. This is usually the case for most seafood products (115).

- Shellfish

Shellfish control and risk management are facilitated by the lack of mobility of the target food product, such as, mussels, clams, and oysters. Therefore, preharvest monitoring and management have the highest potential for effective identification and control of high-risk products. For preharvest monitoring, toxic blooms can be identified and the risk to the consumer avoided by either closing the area to harvesting or, where possible, moving shellfish mariculture rafts out of the area of the toxic blooms. Preharvest monitoring can also include testing the shellfish.

Marketplace monitoring presents another option when preharvest analysis is not possible. For this kind of monitoring program, it is important to establish a good sampling program and to have accurate analytical methodology available. In marketplace sampling, it is important to remember that time is a limiting factor; thus, rapid analytical methodology is essential (Table 7).

A hazard analysis critical control point (HACCP) concept should be considered for seafood safety management programs. The HACCP system provides several magnitudes of food safety assurance over that offered by traditional inspections for food market and food services.

- Finfish

Control and risk management of ciguatoxic finfish before harvesting are hampered by fish mobility. Nevertheless, it is possible to have some control by testing the harvesting areas to determine the toxic potential or identify locations with a prior history of toxicity for causative organisms or toxic fish. High-frequency sampling would then be needed to predict toxic outbreaks. On the other hand, marketplace monitoring along with identification of high-risk fishing locations has the highest potential for effective identification and control of high-risk products. Rapid testing methods, however, are crucial for the adequate control of fish in the market (Table 8).

Generalizations regarding the uptake and retention of phycotoxins by fish or shellfish should be avoided. Differences in rates of toxin accumulation and retention are dependent upon the fish or shellfish and algal species under consideration, and these differences should be taken into account before choosing a species to be reared in areas susceptible to toxic algal blooms (148).

Pufferfish poisoning cannot be prevented by preharvest analysis. Postharvest management and product preparation are crucial to diminish the risk associated with this illness. Adequate training programs for the preparation of fillets must be established in high-risk areas.

The U.S. Food and Drug Administration (FDA) has set up monitoring programs for domestic and imported products. These programs, although set upon a national scale, can be used as an example for other monitoring

Table 7. Overview of seafood monitoring and management programs for risks associated with aquatic biotoxins: shellfish.

Disease	Preharvest		Marketplace		Ease of control/ risk management	High-priority research needs
	Monitoring	**Management**	**Monitoring**	**Management**		
Shellfish:		*				
PSP DSP ASP NSP	Test for presence of causative dinoflagellate(s) or toxin in shellfish in harvesting areas.	+: Identify dinoflagellate species. Test shellfish. Withhold from market until testing results show acceptable levels (depuration).	Collect samples in programmed or random schedule.	+: Close the market. Treat the product or divert to lower risk uses. –: Allow shellfish to enter the marketplace.	**High**: East of preharvest management due to lack of mobility of shellfish.	Monitoring for causative organism(s). Information on ecology of causative organisms.
		If possible, move shellfish (mariculture).				Prediction of blooms.
		–: Allow shellfish to enter the marketplace.				Rapid testing method for marketplace.

PSP, paralytic shellfish poisoning; DSP, diarrheic shellfish poisoning; ASP, amnesic shellfish poisoning; NSP, neurotoxic shellfish poisoning.
*, Highest potential for effective identification and control of high-risk products; +, presence of toxins or dinoflagellates detected at a level of concern; –, presence of toxins or dinoflagellates not detected.

Table 8. Overview of seafood monitoring and management programs for risks associated with aquatic biotoxins: finfish and cyanobacteria.

	Preharvest		Marketplace			
Disease	Monitoring	Management	Monitoring	Management	Ease of control/ risk management	High-priority research needs
Finfish		*	*			
Ciguatera fish poisoning	Test dinoflagellates and/or fish in harvesting areas. Identify biomarker for CFP.	+: Establish sampling programs with greater frequency. –: Allow the fish to enter the marketplace.	Analysis of target fish and fish weighing >5 kg.	+: Close the market. Treat the product or divert to lower risk uses. –: Allow fish to enter the marketplace.	**Low**: Pre-harvest management hampered by mobility of fish. Harvesting area mapping in combination with marketplace management is crucial.	Rapid method testing for marketplace.
Pufferfish poisoning	Not applicable.	Not applicable.	Test fish.	Establish training program for preparation of fillets.	**Moderate.**	Development of educational programs for fish preparation.
Cyanobacteria	Not applicable.	Not applicable.	Analysis of water supply and recreational waters.	+: Use other water supplies. Treat water.	**Moderate.**	Water treatment procedures.

*, Highest potential for effective identification and control of high-risk products; +, presence of toxins or dinoflagellates detected at a level of concern; –, presence of toxins or dinoflagellates not detected.

programs (49, 50). The frequency and number of samples collected for local programs should be increased significantly. Factors affecting the establishment of sampling plans are discussed in Step 1.

- Cyanobacteria

Risks associated with Cyanobacteria need to be addressed in a different manner. Analysis of water supplies for presence of toxin-producing organism(s) is an essential factor for their control. When toxic waters are detected, these should be closed to recreational activities. Water treatment procedures are needed to control water quality and safety (see Table 8).

The availability of laboratory facilities, as well as that of equipment and trained personnel, are also important factors in designing a specific monitoring mechanism for a country or area. In some cases, economic factors such as the budget to acquire analytical equipment will be determining factors. The action of international organizations in aiding developing countries with analytical capabilities is essential. In some cases, training programs are required.

The method intended to be used in a specific seafood safety monitoring program must be standardized. In other words, to be established as the official method of a specific program, it must go through prior validation such as interlaboratory studies to determine precision (reproducibility, repeatability) and accuracy (recovery) parameters of the method (113). If method development capabilities are not available, an official international method should be adopted. In addition to the need to validate methods used in monitoring programs, each laboratory and analyst must demonstrate the capacity to obtain accurate and reliable test results for each analyte and substrate in the monitoring programs. Formal and within-laboratory check sample programs need to be established utilizing reliable analytical standards and standard material. International organizations such as IUPAC and IOC/UNESCO should be contacted to locate sources where these analytical standards and standard material are available.

Step 4: Establish regulatory limits.

Various factors may play a role in establishing regulatory criteria and limits for phycotoxins. These include the availability of survey and toxicological data, the distribution of phycotoxins throughout sampled lots, the stability of the toxins in the samples or test portions, the availability of methods of analysis, the regulations in force in other countries, and enforceability of desired regulations (148, 169). Often with the initial efforts destined to establish regulatory limits for unavoidable natural toxicants, scientists and health regulatory officers are not aware of the complete identity or the relative potencies of the toxins involved with the foodborne diseases. Therefore, it is often necessary to set up a program that will allow the use of an indicator compound to which the remaining compounds could be compared or to monitor an event such as toxic algae blooms (116).

- Shellfish
  A number of countries have established tolerances for PSP, DSP, ASP, and NSP (Table 9) applicable to domestically produced as well as imported shellfish. It is evident that these tolerances vary greatly from country to country. Some of the established regulations may be used as an example for countries establishing their own programs (153).
- Finfish
  For ciguatera, however, the problem of establishing a regulatory monitoring program is compounded even further in that not only is the phenomenon not well understood with regard to which are the toxins involved with the disease and their structural identification, but also the toxins that appear to be of highest potency, the ciguatoxins, are not available in sufficient quantity to be used in a monitoring or regulatory program (11). When this is the case, an alternative compound could be used that takes into account the relative potency of the alternative standard and how it could be used in the establishment of the regulatory screening program (115). Commonly, alternative compounds or standards are used to standardize analytical methods where the toxins have high toxin potentials or limited availability, such as, quinine sulfate for aflatoxins and brevetoxins for ciguatoxins.

Table 9. Summary of regulations for diarrheic shellfish poisons (DSP), paralytic shellfish poisons (PSP), amnesic shellfish poisons (ASP), and neurotoxic shellfish poisons (NSP).

	DSP	PSP	ASP	NSP
Countries	18	25	4	2
Products	Molluscs, shellfish, bivalves, mussels	Molluscs, shellfish, bivalves, mussels	Molluscs, shellfish, bivalves	Shellfish, bivalves
Limits	5 MU/100 g Not detectable by rat bioassay 40–60 μg STX/100 g	400 MU/100 g 40–80 μg PSP/100 g 40–80 μg STX/100 g	2 mg domoic acid/100 g	Not detectable by mouse bioassay
Methods of analysis applied	Mouse bioassay Rat bioassay HPLC	Mouse bioassay Spectrophotometric method HPLC (Sullivan method)	HPLC	Mouse bioassay

HPLC, high performance liquid chromatography; MU, mouse unit; STX, saxitoxin.

Reprinted by permission of CNEVA (France). "Worldwide Regulations for Marine Phycotoxins," CNEVA—Actes du Colloque sur les Biotoxines Marines, by van Egmond, van den Top and Speijers, 1991, pg. 196, and with updates from Shumway et al. 1995.

Step 5: Establish regulatory policy for violative products.

The disposition of product represented by analytical test results should be well established, including permitting acceptable product to enter or continue in commercial channels. The most effective means of controlling quality during outbreaks of toxic algae is either by blanket closure during certain times of the year or by instituting a shellfish toxicity monitoring program (147). When few monitoring stations are employed to cover large coastal areas and the frequency of sampling is relatively low, program officials close large fishery areas as a conservative measure to protect the consumer and maintain consumer confidence in the seafood industry. Monitoring not only is important in the closure but is crucial for establishing reopenings (4).

Authority for handling both known and new marine toxins occurring in seafood is sometimes fragmented within state or national management agencies, leading to inter- and intraagency jurisdictional overlap and confusion. It is important to encourage the locations to streamline programs concerned with marine toxin risks by placing monitoring and management control in a single regulatory agency or establish memoranda of understanding clearly outlining jurisdictional responsabilities (4).

Some of the public health measures that can be adopted include the education of the fishery industry and the public in affected areas about risks associated with the contaminant, closure to fishing of areas known to be highly toxic, bans of the sale of high-risk fish from known toxic locations, detection of toxic fish products before consumption, decontamination treatments, and diversion of toxic products to other uses (1, 139).

Rates of detoxification vary considerably between species, and some species remain toxic for extended periods of time. Transferring large quantities of shellfish to waters free of the toxic organisms for self-depuration is labor intensive and costly. There have been no useful methods devised for effectively reducing phycotoxins in contaminated shellfish. All methods tested to date have been either unsafe, too slow, economically unfeasible, or yielded products unacceptable in appearance and taste (148).

Research has shown that individuals can reduce their risk of contracting ciguatera fish poisoning by the following steps:

- Avoidance of warm water reef fish, particularly those with a known propensity to be toxic, and avoidance of certain pelagic fish that feed on them, especially in areas with a history of ciguatera
- Complete avoidance of moray eels, which are commonly highly toxic (87, 88, 98), except when captured in areas with no history of ciguatera
- Avoidance of carnivorous fish, which may reduce but not eliminate the risk of contracting ciguatera
- Avoidance of the head, roe, and viscera of potentially toxic fish
- Feeding a large fish flesh meal to a cat, which is then observed for at least 6 h before human consumption of portions of the same fish.

Step 6: Distribute information.

When a HAB with toxic potential is encountered, the general public should be warned by immediate distribution of information through the media (television, radio, newspapers) and by means of posters or flyers in harbors, beaches, markets, etc. (2).

Public health officials as well as harvesters, processors, and dealers of shellfish must remain alert to outbreaks of toxic algal blooms to protect human health as well as preserving a high standard of quality assurance (148).

V. Recommendations for Future Activities

Although the potential risks associated with aquatic biotoxins have been well documented, there is an increased need to obtain further information for the development and implementation of an international risk management plan. Aquatic biotoxins are distributed worldwide and have caused a great variety of acute and chronic toxic syndromes.

Risks associated with aquatic biotoxins have been generally identified and evaluated for many regions of the world. The challenge for international research institutes and regulatory agencies will be the development of seafood safety programs that will reduce the occurrence of acute toxicity syndromes and, in the long term, reduce the chronic risk associated with consumption of foods from aquatic resources. The development of monitoring programs that include analytical and decontamination procedures should be an international goal.

Efforts should be made to develop and maintain international collaborative links and information exchanges. Early identification of the toxicological syndrome is necessary for effective therapeutic intervention. Once recognized, proper and prompt reporting can alert official agencies to implement regulatory directives (4). Regulations presently in force in other nations, especially those of trading partners, should be considered when new international regulations are being put into place. Differences between national tolerances set for marine toxins can result in chaos and inconsistencies in the protection of public health (169). A well-defined distribution plan for biotoxin standards isolated or synthesized with the assistance of national or international funding is needed.

A retrospective, collaborative survey of historical phytoplankton data on the spatial distribution of toxic algae, bacteria, and cyanobacteria should be conducted worldwide. Efforts to determine alternative stages in the life history of toxic species, including possible resting stages, should be expanded to include both cultured isolates and natural populations. On the other hand, there is serious concern for hitherto unrecorded toxins, such as the hallucinogenic toxins reported in Hawaii, which highlight the need to maintain a worldwide watching brief.

It is necessary to develop a database on seafood intoxications similar to that developed by the U.S. Center for Disease Control and Prevention (CDC). Also, it is important to explore the development of better reporting tools. It is also necessary to educate physicians, public health officials, and consumers in issues of seafood poisoning (4).

There is a need for the development of standardized, generic laboratory procedures, techniques, protocols, and surveys that could be compiled and be "ready-to-use" by local and state public health agencies (4, 75). Priority should be given to the development and implementation of reliable chemical analytical methods and assays to reduce dependence on the mouse bioassay. Control measures should be applied initially at the earliest stage of seafood production by monitoring of water quality and condition (4, 114). Rapid and simple tests should be developed and used to screen potentially hazardous fish or shellfish at the point of harvest to reduce costs to the fishermen and to protect the consumer from toxins and dangerous contaminants (4). New or improved methodologies (immunological methods, gene probe, polymerase chain reaction) should be developed to provide for the rapid identification and quantification of indicators, seafood-associated pathogens, and microbial toxins in seafood and harvest waters (4).

National and international guidelines for microbial and natural toxin contamination should be extended and updated on a routine basis. There is evidence that existing guidelines have not been adequately conveyed to the fishing industry or to interested members of the public. Regulatory agencies should develop a set of monitoring and inspection practices focusing more strongly on environmental conditions and on contaminant levels in the edible portion of seafood at the point of capture (4). Programs should be established for training regulators and seafood industry personnel to be proficient in the regulatory programs under consideration. Educational programs for safe preparation and service of seafood in commercial and homesettings must also be developed and delivered as a part of an integrated seafood safety program.

A. Recommendations for the Development of Biotoxin Detection Methods

1. Develop and validate methods for the detection of biotoxins in toxigenic algal species, seafoods, and environmental samples.
2. Develop and validate rapid, low-cost immunochemical methods as preemptive (screening) detection tools.
3. Continue with the development and improvement of analytical methods for confirmation of identity determinations from biological and environmental samples.
4. Identify bioindicators for toxic potential in fish and shellfish harvesting areas.

B. Recommendations on the Biology of the Biotoxin Producers

1. Determine the occurrence and distribution of the known and potentially toxic microbial species, both motile and encysted.
2. Identify poorly characterized aquatic biotoxins from known and suspected toxigenic algae, bacteria, and cyanobacteria.

3. Determine the quantitative degrees of biotoxin production, and biotoxin distribution in fish and shellfish tissues.
4. Determine the natural or anthropogenic factors that promote the growth and dispersion of biotoxin producers.
5. Increase field and surveillance studies, which should include the improvement of remote sensing technologies for detecting blooms at early stages of development. Incorporate the use of conventional field sampling and monitoring by seagoing vessels with improved and standardized methods for the detection and characterization of motile populations and cyst beds into surveillance programs.
6. Conduct laboratory studies on the physiological ecology, biotoxin production, and potency in a cross section of clonal, population, and cyst isolates as well as spatial and temporal isolates of biotoxin producers.
7. Determine the physical factors involved in the initiation, propagation, transport, and termination of blooms.

C. Recommendations for Public Health Concerns

1. Determine the extent and potential for acute and chronic impact on human health.
2. Determine the modes of toxin impact on human health and the probabilities of symptomatic-making, incomplete, or misdiagnoses.
3. Provide an epidemiological assessment of seafood poisonings and the development ofremedial medical treatments.
4. Develop and validate diagnostic methods for aquatic biotoxin-related illnesses.
5. Develop antidotes and treatment procedures for victims of poisoning outbreaks.

D. Recommendations on the Development of Seafood Safety Monitoring Programs

1. Determine principal agents or organisms responsible for the production of each class of aquatic biotoxins.
2. Evaluate relative toxic potencies for aquatic biotoxins within each group.
3. Identify major seafood products associated with toxin accumulation and/or retention.
4. Identify critical control points within the fishing industry operations for each class of toxin.
5. Develop and validate analytical tools, i.e., screening methods, confirmation of identity, etc., for use at critical control points such as harvesting areas, aboard fishing vessels, receiving docks, processing plants, distribution and retail outlets, and regulatory laboratories.
6. Develop decontamination procedures or diversion for lower-risk uses of contaminated products.

E. Future Research Focus

1. Provide an adequate source of purified toxins for the provision of analytical standards, method development and validation, therapeutic treatment, and biological evaluation.
2. Continue with the development of analytical methods for the rapid low-level detection of the toxins and confirmation of identity.
3. Conduct animal studies, including the effect of chronic exposure to low doses of the toxins, to determine the potential for long-term adverse health effects.
4. Continue with the studies of the factors that control the synthesis of the toxins by dinoflagellates, bacteria, and cyanobacteria.
5. Study the ecology of toxigenic strains of dinoflagellates to determine their growth characteristics, environmental conditions affecting blooms, and eventual fate of the biotoxins in seawater.
6. Develop test kits for rapid, affordable detection of toxic outbreaks.
7. Monitor fish and shellfish for new toxic agents.

Summary

Naturally occurring toxicants are usually odorless, tasteless, and generally undetectable by any simple chemical test. Various programs have been established that are effective in reducing risks associated with these toxicants in food. These programs include setting regulatory limits, monitoring susceptible commodities for toxin levels, and using decontamination procedures. Bioassays have been used traditionally to monitor suspect products. All traditional bioassays, however, have one common disadvantage, i.e., the lack of specificity for individual toxins. The lack of available reference standards for specific toxins has also hampered implementation of monitoring programs. Utilizing the knowledge gained with regulatory monitoring and decontamination programs for other toxins, e.g., aflatoxin, similar seafood safety programs can be developed for aquatic biotoxins that will reduce risks and hazards associated with the contaminant to practicable levels and help to preserve an adequate food supply.

Research is needed in several areas identified in this article. International cooperation has an important role in achieving these essential elements. Global programs will help in the adequate management of risks associated with aquatic biotoxins. To have an effective monitoring program, it is necessary to define precisely the local needs for information in a short or long time range. It is necessary to have basic knowledge about the biological, chemical, and physical conditions as well as temporal and geographic variations within the region of interest (2). Regardless of the overall success of fish/shellfish toxin monitoring plans, emergencies will occur. Therefore, contingency plans should be developed so there will be no misunderstanding of what actions to take (148). In general, however, the structure of the program must be kept as simple as possible to facilitate fast and uncomplicated flow of information among the various organizations and individuals involved (2).

Public health and safety requires the removal of any toxic shellfish from the market, within practicability, and closure of any suspect harvest area. It should be important to remember that economic value of the fish or shellfish resource is always secondary to public health and safety (148).

References

1. Ahmed FE (1991) Naturally occurring fish and shellfish poisons. In: Ahmed FE (ed) Seafood Safety. National Academy Press, Washington, DC, pp 87–110.
2. Andersen P (1996) Design and implementation of some harmful algal monitoring systems. UNESCO, Paris, France.
3. Anderson DM (1989) Toxic algal blooms and red tides: a global perspective. In: Okaichi T, Anderson DM, Nemoto T (eds) Red Tides: Biology, Environmental Science and Toxicology. Elsevier, New York.
4. Anderson DM, Galloway SB, Joseph JD (1993) Marine biotoxins and harmful algae: a national plan. Technical report. Woods Hole Oceanographic Institution.
5. Anderson DM (1994) Red tides. Sci Am 271:52.
6. AOAC (1990) Paralytic shellfish poison. Biological method. Final action. Method 959.08. In: Hellrich K (ed) Official Methods of Analysis. AOAC International, Arlington, VA, pp 881–882.
7. Ayer SW, Pleasance S, Laycock MV, Thibault P (1992) Ionspray mass spectrometry of marine toxins: analysis of paralytic shellfish poisoning toxins by flow injection, LC-MS, and CE-MS. In: Therriault JC, Levasseur M (eds) Proceedings of the 3rd Canadian Workshop on Harmful Marine Algae, Mont-Joli, Quebec, p 15.
8. Baden DG, Rein KS, Kinoshita M, Gawley RE (1990) Computational modeling of the polyether ladder toxins brevetoxin and ciguatoxin. In: Tosteson TR (ed) Proceedings of the 3rd International Conference on Ciguatera Fish Poisoning. Polyscience Publications, Quebec, pp 103–114.
9. Baden DG, Sechet VM, Rein KS, Edwards RA (1992) Methods for detecting brevetoxins in seawater, in biological matrices revealed, and on excitable membranes. Bull Soc Pathol Ex. 85:516–519.
10. Bagnis R, Kuberski T, Laugier S (1979) Clinical observations on 3,009 cases of ciguatera (fish poisoning) in the South Pacific. Am J Trop Med Hyg 28:1067–1073.
11. Bagnis R, Barsinas M, Prieur C, Pompon A, Chungue E, Legrand AM (1987) The use of mosquito bioassay for determining toxicity to man of ciguateric fish. Biol Bull 172:137–143.
12. Bates SS, Bird CJ, Boyd RK, De Freitas ASW, Flak M, Foxall RA, Hanic LA, Jamieson WD, McCulloch AW, Odense P, Quilliam MA, Sim PG, Thibault P, Walter JA, Wright JLC (1988) Investigations on the source of domoic acid responsible for the outbreak of amnesic shellfish poisoning (ASP) in Eastern Prince Edward Island. Tech Rep 57. Atlantic Research Laboratory, National Research Council, Halifax, Nova Scotia.
13. Beltrami E, Copper E (1991) Modeling the temporal dynamics of unusual blooms. Presented at the Fifth International Conference on Toxic Marine Phytoplankton, 28 Oct–1 Nov, Newport, RI, U.S.A.
14. Blay P, Boyd RK, Janecek M, Kelly J, Locke S, Pleasance S, Quilliam MA, Thi-

bault P (1992) Development of instrumental analytical methods for the determination of saxitoxin and tetrodotoxin in water and soil. IMB Tech Rep 66:1–88.
15. Blythe DG, Fleming LE, Ayar DR, De Sylva D, Baden DG, Schranik K (1994) Mannitol therapy for acute and chronic ciguatera fish poisoning. Mem Queensl Mus 34:465–470.
16. Botes DP, Kruger H, Viljoen CC (1982) Isolation and characterization of four toxins from the blue-green alga, *Microcystis aeruginosa*. Toxicon 20(6):945–954.
17. Botes DP, Viljoen CC, Kruger H, Wessts PL, Williams DH (1982b) Structure of toxins of the blue-green alga *Microcystis aeruginosa*. S Afr J Sci 78:378–379.
18. Bouaicha N, Hennion MC, Sandra P (1997) Determination of okadaic acid by micellar electrokinetic chromatography with ultraviolet detection. Toxicon 35:273.
19. Brooks WP, Codd GA (1988) Immunoassay of hepatotoxic cultures and water blooms of cyanobacteria using *Microcystis aeruginosa* peptide toxin polyclonal antibodies. Environ Tech Lett 9:1343–1348.
20. Buckley LJ, Ikawa M, Sasner JJ (1976) Isolation of *Gonyaulax tamarensis* toxins from soft shell clams and a thin-layer chromatographic-fluorimetric method for their detection. J Agric Food Chem 24:107–111.
21. Canet C (1993) Importance of international cooperation in food safety. Food Addit Contam 10:97–104.
22. Carlson RE, Lever ML, Lee BW, Guire PE (1984) Development of immunoassays for paralytic shellfish poisoning. In: Ragelis EP (ed) Seafood Toxins. American Chemical Society, Washington, DC, pp 181–192.
23. Carmichael WW, Mahmood NA (1984) Toxins from freshwater cyanobacteria. In: Ragelis EP (ed) Seafood Toxins. American Chemical Society, Washington, DC, pp 87–106.
24. Carmichael WW, Mahmood NA, Hyde EG (1990) Natural toxins from cyanobacteria (blue-green algae). In: Hall S, Strichartz G (eds) Marine Toxins: Origin, Structure, and Molecular Pharmacology. American Chemical Society, Washington, DC, pp 87–106.
25. Carmichael WW (1992) Cyanobacteria secondary metabolites—the cyanotoxins. J Appl Bacteriol 72:445–459.
26. Carmichael WW, Falconer IR (1993) Diseases related to freshwater blue-green algal toxins, and control measures. In: Falconer IR (ed) Algal Toxins in Seafood and Drinking Water. Academic Press, London, pp 187–209.
27. Carmichael WW (1994) The toxins of cyanobacteria. Sci Am 270(1):78–86.
28. Carmody EP, James KJ, Kelly SS (1995) Diarrhetic shellfish poisoning: evaluation of enzyme-linked immunosorbent assay methods for the determination of dinophysistoxin-2. J AOAC Int 78(6):1403–1408.
29. Carver CE, Hancock S, Sims GG, Watson-Wright W (1992) Phytoplankton monitoring in Nova Scotia. In: Therriault JC, Levasseur ML (eds) Proceedings of the Third Canadian Workshop on Harmful Marine Algae, Mont-Joli, Quebec, p 18.
30. Catterall WA, Gainer M (1985) Interaction of brevetoxin A with a new receptor site on the sodium channel. Toxicon 23(3):497–504.
31. Cembella AD, Lamoureux G (1993) A competitive inhibition enzyme-linked immunoassay for the detection of paralytic shellfish toxins in marine phytoplankton. In: Smayda TJ, Shimizu Y (eds) Toxic Phytoplankton Blooms in the Sea. Elsevier, Amsterdam, pp 857–862.
32. Cembella AD, Todd E (1993) Seafood toxins of algal origin and their control in

Canada. In: Falconer IR (ed) Algal Toxins in Seafood and Drinking Water. Academic Press, London, pp 129–144.

33. Cembella AD, Milenkovic LV, Doucette GJ (1995) Part A. *In vitro* biochemical and cellular assays. In: Hallegraeff GM, Anderson DM, Cembella AD (eds) Manual on Harmful Marine Microalgae. UNESCO, Paris, France, pp 181–212.
34. Chu FS, Huang X, Wei RD, Carmichael WW (1989) Production and characterization of antibodies against microcystins. Appl Environ Microbiol 55(8):1928–1933.
35. Chu FS, Huang X, Wei RD (1990) Enzyme-linked immunosorbent assay for microcystins in blue-green algal blooms. J AOAC 73(3):451–456.
36. Chungue E, Bagnis RA, Parc E (1984) The use of mosquito (*Aedes aegypti*) to detect ciguatoxin in surgeon fishes (*Ctenochaetus striatus*). Toxicon 22:161.
37. Croci L, Cozzi L, Stacchini A, De Medici D, Toti L (1997) A rapid tissue culture assay for the detection of okadaic acid and related compounds in mussels. Toxicon 35:223.
38. Dallinga-Hanneman L, Liebezeit G, Zeeck E (1993) Development of fast and sensitive tests for the toxic non-protein aminoacids, kainic acid and domoic acid. Presented at the 6th International Conference on Toxic Phytoplankton, Oct 18–22, Nantes, France.
39. Dalzell P (1994) Management of ciguatera fish poisoning in the South Pacific. Mem Queensl Mus 34(3):471–479.
40. Della Loggia R, Cabrini M, Del Negro P, Honsell G, Tubaro A (1993) Relationship between *Dinophysis* spp. in seawater and DSP toxins in mussels in the Northern Adriatic Sea. In: Smayda TJ, Shimizu Y (eds) Toxic Phytoplankton Blooms in the Sea. Elsevier, New York, pp 483–488.
41. Del Negro P, Cabrini M, Tulli F (1993) A comparative analysis of toxic algae in *Mytilus galloprovincialis* and in *Crassostrea gigas*. In: World Aquaculture. Special Publ 19:42. European Aquaculture Society, Belgium.
42. Del Negro P, Cabrini M, Tulli F (1993) Toxic dinoflagellates in mussels and in seawater: a new assessment in shellfish farming water quality evaluation. In: Bamabe G, Kestemont P (eds) Production, Environment, and Quality. Special Publ No. 18. European Aquaculture Society, Belgium, pp 555–562.
43. Dickey R (1989) Marine biotoxins and associated public health risks in the Gulf of Mexico. In: Burrage D (ed) Proceedings of a Mississippi Sea Grant Advisory Service Workshop. Mississippi Sea Grant Advisory Service, Mississippi Cooperative Extension Service, Mississippi State University, Mississippi/Alabama Sea Grant Consortium, and U.S. Environmental Protection Agency Gulf of Mexico Program, pp 15–21.
44. Dickey RW, Bencsath FA, Granade HR, Lewis RJ (1992) Liquid chromatographic-mass spectrometric methods for the determination of marine polyether toxins. Bull Soc Pathol Ex. 85:514–515.
45. Do HK, Hamasaki K, Simidu U, Noguchi T, Shida Y, Kogune K (1993) Presence of tetrodotoxin and tetrodotoxin-producing bacteria in freshwater sediments. Appl Environ Microbiol 59(11):3934–3937.
46. Doorenbos NJ (1984) Ciguatera toxins: where do we go from here? In: Ragelis EP (ed) Seafood Toxins. American Chemical Society, Washington, DC, pp 60–73.
47. Elbusto C, Carreto JI, Benavides HR, Sancho H, Carignan MO, Oshima Y, Yasumoto T (1991) Paralytic shellfish toxin profiles in the marine snail, *Zidona angulata*, from the Mar del Plata coast. Presented at the Fifth International Conference on Toxic Marine Phytoplankton, 28 Oct–1 Nov, Newport, RI, U.S.A.

48. Elbusto C, Carreto JI, Benavides HR, Sancho H, Cucchi Colleoni D, Carignan MO, Fernandez A (1991) Paralytic shellfish toxicity in the Argentine Sea, 1990: an extraordinary year. Presented at the Fifth International Conference on Toxic Marine Phytoplankton, 28 Oct–1 Nov, Newport, RI, U.S.A.
49. FDA (1996) Compliance Program 7303.842. Domestic fish and fishery products inspection program. Project 07: Natural poisons. U.S. Food and Drug Administration, Center for Food Safety and Applied Nutrition, Office of Seafood, Washington, DC.
50. FDA (1996) Compliance Program 7303.844. Import seafood product program. Project 07: Molecular biology and natural toxins. U.S. Food and Drug Administration, Center for Food Safety and Applied Nutrition, Office of Seafood, Washington, DC.
51. Fremy JM, Park DL, Gleizes E, Mohapatra SK, Goldsmith CH, Sikorska HM (1994) Application of immunochemical methods for the detection of okadaic acid in mussels. J Nat Toxins 3(2):95–105.
52. Fritz L, Quilliam MA, Wright JLC, Beale AM, Work TM (1992) An outbreak of domoic acid poisoning attributed to the pennate diatom *Pseudonitzchia australis*. J Phycol 28:439–442.
53. Gathercole PS, Thiel PG (1987) Liquid chromatographic determination of the cyanoginosis, toxins produced by the cyanobacterium *Microcystis aeruginosa*. J Chromatogr 408:435–440.
54. Gillespie NC, Lewis RJ, Pearn JH, Bourke ATC, Holmes MJ, Bourke JB, Shields WJ (1986) Ciguatera in Australia: occurrence, clinical features, pathophysiology and management. Med J Aust 145:584–590.
55. Grimmelt B, Nijjar MS, Brown J, MacNair N, Wagner S, Johnson GR, Ahmend JF (1990) Relationship between domoic acid levels in the blue mussel (*Mytilus edulis*) and toxicity in mice. Toxicon 28:501–508.
56. Hald B, Bjergskov T, Emsholm H (1991) Monitoring and analytical programmes on phycotoxins in Denmark. In: Fremy JM (ed) Proceedings of Symposium on Marine Biotoxins. CNEVA, France, pp 181–187.
57. Hall RL (1988) Food safety in the year 2000. Presented at the IFT/AMA Conference, March, Washington, DC.
58. Hall RL (1991) Toxicological burdens and the shifting burden of toxicology. Presented at the Annual Meeting of the Institute of Food Technologists, June 5, Dallas, TX, U.S.A.
59. Hall S, Strichartz G, Moczydlowski E, Ravindran A, Reichardt PB (1990) The saxitoxins. Sources, chemistry, and pharmacology. In: Hall S, Strichartz G (eds) Marine Toxins: Origin, Structure, and Molecular Pharmacology. American Chemical Society, Washington, DC, pp 29–65.
60. Hallegraeff GM, Bolch CJ (1991) Transport of toxic dinoflagellate cysts in ships' ballast water. Presented at the Fifth International Conference on Toxic Marine Phytoplankton, Oct 28–Nov 1, Newport, RI, U.S.A.
61. Hallegraeff GM (1993) A review of harmful algal blooms and their apparent global increase. Phycologia 32:79–99.
62. Hallegraeff GM (1995) Harmful algal blooms: a global overview. In: Hallegraeff GM, Anderson DM, Cembella AD (eds) Manual on Harmful Marine Microalgae. UNESCO, Paris, France, pp 1–24.
63. Halstead BW (1967) Poisonous and Venomous Marine Animals of the World. Vol 2. Vertebrates. U.S. Government Printing Office, Washington, DC.
64. Harada T, Oshima Y, Yasumoto T (1982) Structures of two paralytic shellfish tox-

ins, gonyautoxins V and VI, isolated from a tropical dinoflagellate, *Pyrodinium bahamense* var. *compressa.* Agric Biol Chem 46:1861–1864.

65. Harada KI, Ogawa K, Matsuura K, Nagai H, Murata H, Suzuki M, Itezuno Y, Nakayama N, Shirar M, Nakano M (1991) Isolation of two toxic heptapeptide microcystins from an axenic strain of *Microcystis aeruginosa,* K-139. Toxicon 29(4/5):479–489.
66. Hoffman PA, Granade HR, McMillan JP (1983) The mouse ciguatoxin bioassay: a dose response curve and symptomatology analysis. Toxicon 21(3):363–369.
67. Hokama Y, Banner AH, Boyland DB (1977) A radioimmunoassay for the detection of ciguatoxin. Toxicon 15:317.
68. Hokama Y (1985) A rapid, simplified enzyme immunoassay stick test for the detection of ciguatoxin and related polyethers from fish tissues. Toxicon 23:939–946.
69. Hokama Y, Miyahara JT (1986) Ciguatera poisoning: clinical and immunological aspects. J Toxicol Toxin Rev 5:23–25.
70. Hokama Y (1993) Recent methods for detection of seafood toxins: recent immunological methods for ciguatoxin and related polyethers. Food Addit Contam 10(1): 71–82.
71. Holmes CFB (1991) Liquid chromatography-linked protein phosphatase bioassay: a highly sensitive marine bioscreen for okadaic acid and related diarrhetic shellfish toxins. Toxicon 29:469–477.
72. Holmes MJ, Gillespie NC, Lewis RJ (1988) Toxicity and morphology of *Ostreopsis* cf. *siamensis* cultured from a ciguatera endemic region of Queensland, Australia. In: Proceedings of the Sixth International Coral Reef Symposium, vol 3, pp 49–54.
73. Honkanen RE, Mowdy DE, Dickey RW (1996) Detection of DSP toxins, okadaic acid and dinophysistoxin-1 in shellfish by serine/threonine protein phosphatase assay. J AOAC Int 79(6):1336–1343.
74. Honsell G, Boni L, Cabrini M, Pompei M (1992) Toxic or potentially toxic dinoflagellates from the Northern Adriatic Sea. In: Science of the Total Environment. Elsevier, Amsterdam, pp 107–114.
75. Hungerford JM, Lee S, Hall S (1991) A rapid screening method for PSP toxins in shellfish based on flow injection analysis. Presented at the Fifth International Conference on Toxic Marine Phytoplankton, Oct 28–Nov 1, Newport, RI, U.S.A.
76. International Programme on Chemical Safety and International Life Sciences Institute, Europe (1992) Final report of the steering group on naturally occurring toxins of plant origin. Carshalton, UK.
77. Kirimura LH, Abad MA, Hokama Y (1982) Evaluation of the radioimmunoassay (RIA) for detection of ciguatoxin (CTX) in fish tissues. J Fish Biol 21:671.
78. Lassus P (1994) Sixth international conference on toxic marine phytoplankton. Harmful Algae News 8:3.
79. Lawrence JF (1990) Determination of domoic acid in seafoods and in biological tissues and fluids. In: Hynie I, Todd ECD (eds) Proceedings of a symposium on domoic acid toxicity, Ottawa, Ontario, pp 27–31.
80. Lawrence JF, Menard C (1991) Determination of paralytic shellfish poisons by prechromatographic oxidation and HPLC. In: Fremy JM (ed) Proceedings of symposium on marine biotoxins. CNEVA, Paris, pp 127–130.
81. Lawrence JF, Wong B, Menard C (1995) Determination of decarbamoyl saxitoxin and its analogues in shellfish by prechromatographic oxidation and liquid chromatography with fluorescence detection. J AOAC Int 78:1111–1115.

82. Ledoux M, Fremy JM (1994) Phytoplancton, phycotoxines et intoxications alimentaires. Rec Med Vet 170:129–139.
83. Lee JS, Yanaga T, Kenma R, Yasumoto T (1987) Fluorometric determination of diarrhetic shellfish toxins by high performance liquid chromatography. Agric Biol Chem 51:877–881.
84. Lee JS, Igarashi T, Fraga S, Dahl E, Hovgaard P, Yasumoto T (1989) Determination of diarrhetic shellfish toxins in various dinoflagellate species. J Appl Phycol 1: 147–152.
85. Legrand AM, Fukui M, Cruchet P, Yasumoto T (1992) Progress on chemical knowledge of ciguatoxins. Bull Soc Pathol Ex. 85:467–469.
86. Levine L, Fujiki H, Yamada K, Ojika M, Gjika HB, van Vunakis H (1988) Production of antibodies and development of a radioimmunoassay for okadaic acid. Toxicon 26:1123–1128.
87. Lewis RJ, Sellin M, Poli MA, Norton RS, MacLeod JK, Sheil MM (1991) Purification and characterization of ciguatoxins from moray eel (*Lycodontis javanicus*, Muraenidae). Toxicon 29(9):1115–1127.
88. Lewis RJ, Sellin M, Street R, Holmes MH, Gillespie NC (1992) Excretion of ciguatoxin from moray eels (Muraenidae) of the Central Pacific. In: Tosteson TR (ed) Proceedings of the 3rd International Conference on Ciguatera Fish Poisoning. Polyscience Publications, Quebec, pp 131–143.
89. Lewis RJ, Holmes MJ (1993) Origin and transfer of toxins involved in ciguatera. Comp Biochem Physiol 106C(3):615–628.
90. Lewis RJ, Sellin M (1993) Recovery of ciguatoxin from fish flesh. Toxicon 31(10): 1333–1336.
91. Lewis RJ (1994) Impact of validated, cost-effective screen for ciguateric fish. Mem Queensl Mus 34(3):549–553.
92. Lewis RJ, Holmes MJ, Alewood PF, Jones A (1994) Ionspray mass spectrometry of ciguatoxin 1, maitotoxin-2 and -3, and related marine polyether toxins. Nat Toxins 2:56–63.
93. Lewis RJ (1995) Detection of ciguatoxins and related benthic dinoflagellate toxins: *in vivo* and *in vitro* methods. In: Hallegraeff GM, Anderson DM, Cembella AD (eds) Manual on Harmful Marine Microalgae. UNESCO, Paris, France, pp 135–162.
94. Manger RL, Leja LS, Hungerford JM, Wekell MM (1994) Cell bioassay for the detection of ciguatoxins, brevetoxins, and saxitoxins. Mem Queensl Mus 34(3):571–575.
95. Manger RL, Leja LS, Hungerford JM, Hokama Y, Dickey RW, Granade HR, Lewis R, Yasumoto T, Wekell MM (1995) Detection of sodium channel toxins: directed cytotoxicity assays of purified ciguatoxins, brevetoxins, and seafood extracts. J AOAC Int 78(2):521–527.
96. Moore RE (1984) Public health and toxins from marine blue-green algae. In: Ragelis EP (ed) Seafood Toxins. American Chemical Society, Washington, DC, pp 369–376.
97. Mosher HS, Fuhrman FA (1984) Occurrence and origin of tetrodotoxin. In: Ragelis EP (ed) Seafood Toxins. American Chemical Society, Washington, DC, pp 333–344.
98. Murata M, Legrand AM, Ishibashi Y, Fukui M, Yasumoto T (1990) Structures and configurations of ciguatoxin from the moray eel *Gymnothorax javanicus* and its likely precursor from the dinoflagellate *Gambierdiscus toxicus*. J Am Chem Soc 112:4380–4386.

99. Murata M, Naoki H, Iwashita T, Matsunaga S, Sasaki M, Yokoyama A, Yasumoto T (1993) Structure of maitotoxin. J Am Chem Soc 115:2060–2062.
100. Murphy M (1993) Plankton monitoring vital for the future of aquaculture! World Aquacult 24: 20–25.
101. Namikoshi M, Rinehart KL, Sakai R, Sivonen K, Carmichael WW (1990) Structures of three new cyclic heptapeptide hepatotoxins produced by the cyanobacterium (blue-green algae) *Nostoc* sp. strain 152. J Org Chem 55:6135–6139.
102. Newsome H, Truelove J, Hierlihy L, Collins P (1991) Determination of domoic acid in serum and urine by immunochemical analysis. Bull Environ Contam Toxicol 47:329–334.
103. Nguyen AL, Luong JH, Massoc C (1990) Capillary electrophoresis for detection and quantitation of domoic acid in mussels. Anal Lett 23:1621–1634.
104. Official Journal of the European Communities (1991) Council Directive of July 15, No L 268/1.
105. Ogura Y (1971) Fugu (puffer-fish) poisoning and the pharmacology of crystalline tetrodotoxin in poisoning. In: Simpson LL (ed) Neuropoisons. Plenum Press, New York, pp 139–158.
106. Onoue Y, Noguchi T, Hashimoto K (1984) Tetrodotoxin determination methods. In: Ragelis EP (ed) Seafood Toxins. American Chemical Society, Washington, DC, pp 345–355.
107. Oshima Y, Kotaki Y, Harada T, Yasumoto T (1984) Paralytic shellfish toxins in tropical waters. In: Ragelis EP (ed) Seafood Toxins. American Chemical Society, Washington, DC, pp 161–170.
108. Oshima Y (1995) Post-column derivatization HPLC methods for paralytic shellfish poisons. In: Hallegraeff GM, Anderson DM, Cembella AD (eds) Manual on Harmful Marine Microalgae. UNESCO, Paris, France, pp 81–94.
109. Paez de Leon L (1985) Aspectos epidemiologicos sobre la intoxicacion paralitica por mariscos. Vet Trop 10:59–85.
110. Palafox NA, Jain LG, Pinano AZ, Gulick TM, Williams RK, Schatz IJ (1988) Successful treatment of ciguatera fish poisoning with intravenous mannitol. JAMA 259: 2740–2742.
111. Park DL, Stoloff L (1989) Aflatoxin control—how a regulatory agency managed risk from an unavoidable natural toxicant in food and feed. Regul Toxicol Pharmacol 9:109–130.
112. Park DL (1993) Prediction of aquatic biotoxin potential in fish and shellfish harvesting areas: ciguatera and diarrheic shellfish poisoning. In: Croce DM, Connell S, Abel R (eds) Coastal Ocean Space Utilization III. Chapman & Hall, London, pp 271–282.
113. Park DL (1994) Reef management and seafood monitoring programs for ciguatera. Mem Queensl Mus 34(3):587–594.
114. Park DL (1995) Detection of ciguatera and diarrheic shellfish toxins in finfish and shellfish with Ciguatect kit. J AOAC Int 78(2):533–537.
115. Park DL (1995) Surveillance programmes for managing risks from naturally occurring toxins. Food Addit Contam 12:623–633.
116. Pearn J (1994) Ciguatera: dilemmas in clinical recognition, presentation and management. Mem Queensl Mus 34(3):601–604.
117. Pereira A, Klein D, Sohet K, Houvenaghel G, Braekman JC (1993) Improved HPLC analysis method for the determination of acidic DSP toxins. Presented at the Sixth

International Conference on Toxic Marine Phytoplankton, Oct 18–22, Nantes, France.

118. Pleasance S, Xie M, LeBlanc Y, Quilliam MA (1990) Analysis of domoic acid and related compounds by mass spectrometry and gas chromatography/mass spectrometry as *N*-trifluoroacetyl-*O*-silyl derivatives. Biomed Environ Mass Spectrom 19: 420–427.
119. Pleasance S, Douglas D, de Freitas ASW, Fritz L, Gilgan MW, Hu T, Marr JC, Quilliam MA, Smyth C, Walter J, Wright JLC (1991) Confirmation of an incident of diarrhetic shellfish poisoning in North America by combined liquid chromatography with ionspray mass spectrometry. In: Proccedings, 39th ASMS Conference, Nashville, TN, pp 1655–1656.
120. Pocklington R, Milley JE, Bates SS, Bird CJ, de Freitas ASW, Quilliam MA (1990) Trace determination of domoic acid in seawater and plankton by high-performance liquid chromatography of the flourenulmethoxycarbonyl (FMOC) derivative. Int J Environ Anal Chem 38:351–368.
121. Poli MA, Templeton CB, Pace JG, Hines HB (1990) Detection, metabolism, and pathophysiology of brevetoxins. In: Hall S, Strichartz, G (eds) Marine Toxins: Origin, Structure, and Molecular Pharmacology. American Chemical Society, Washington, DC, pp 176–191.
122. Poli MA, Hewetson JF (1992) Antibody production and development of a radioimmunoassay for the PbTx-2-type brevetoxins. In: Tosteson TR (ed) Proceedings of the Third International Conference on Ciguatera Fish Poisoning. Polyscience Publications, Quebec, pp 115–127.
123. Prakash A, Medrof JC, Tennant AD (1971) Paralytic shellfish poisoning in Eastern Canada. Fish Res Board Can Bull 177:87.
124. Premazzi G, Volterra L (1993) Microphyte toxins: a manual for toxin detection, environmental monitoring and therapies to counteract intoxications. Environment Institute, Joint Research Centre, Commission of the European Communities, Luxembourg.
125. Quilliam MA, Sim PG, McCulloch AW, McInnes AG (1988) Determination of domoic acid in shellfish tissue by HPLC. Tech Rep #55; NRCC #29015. Atlantic Research Laboratory, National Research Council, Halifax, Nova Scotia.
126. Quilliam MA, Thomson BA, Scott GJ, Siu KWM (1989) Ion-spray mass spectrometry of marine neurotoxins. Rapid Commun Mass Spectrom 3:145–150.
127. Quilliam MA, Pleasance S (1991) Liquid chromatography/mass spectrometry for the analysis of marine toxins. In: Fremy JM (ed) Proceedings on Symposium on Marine Biotoxins. CNEVA, Paris, pp 131–136.
128. Quilliam MA, Hardstaff WR, Marr JC, McDowell LS, Pleasance S (1991) Recent developments in instrumental analytical methods for DSP toxins. Presented at the Fifth International Conference on Toxic Marine Phytoplankton, 28 Oct–1 Nov, Newport, RI, U.S.A.
129. Quilliam MA, Ayer SW, Pleasance S, Sim PG, Thibault P, Marr JC (1992) Recent developments in instrumental analytical methods for marine toxins. In: Bligh EG (ed) Seafood Science and Technology. Fishing Boo News, Blackwell, Oxford, pp 376–386.
130. Quilliam MA (1995) Analysis of diarrhetic shellfish poisoning toxins in shellfish tissue by liquid chromatography with fluorometric and mass spectrometric detection. J AOAC Int 78:555–570.

131. Quilliam MA, Wright JLC (1995) Methods for diarrhetic shellfish poisons. In: Hallegraeff GM, Anderson DM, Cembella AD (eds) Manual on Harmful Marine Microalgae. UNESCO, Paris, France, pp 95–112.
132. Quilliam MA, Xie M, Hardstaff WR (1995) Rapid extraction and cleanup for liquid chromatographic determination of domoic acid in unsalted seafood. J AOAC Int 78: 543–554.
133. Quilliam MA, Hardstaff WR, Ishida N, McLachlan JL, Reeves AR, Ross NW, Windust AJ (1996) Production of diarrhetic shellfish poisoning (DSP) toxins by *Prorocentrum lima* in culture and development of analytical methods. In: Yasumoto T, Oshima Y, Fukuyo Y (eds) Harmful and Toxic Algal Blooms. IOC, UNESCO, Japan, pp 289–292.
134. Ragelis EP (1984) Ciguatera seafood poisoning. In: Ragelis EP (ed) Seafood Toxins. American Chemical Society, Washington, DC, pp 25–36.
135. Ravn H (1995) HAB Publication Series, Vol. 1. Amnesic Shellfish Poisoning (ASP). IOC Manuals and Guides 31(1). UNESCO, Paris, France, pp. 1–15.
136. Ressom R, San Soong F, Fitzgerald J, Turczynowicz L, El Saadi O, Roder D, Maynard T, Falconer I (1994) Health effects of toxic cyanobacteria (blue-green algae). National Health and Medical Research Council,
137. Rigby GR, Steverson IG, Hallegraeff GM (1991) The transfer and treatment of shipping ballast waters to reduce the dispersal of toxic marine dinoflagellates. Presented at the Fifth International Conference on Toxic Marine Phytoplankton, Oct 28–Nov 1, Newport, RI, U.S.A.
138. Rinehart KL, Harada KI, Namikoshi CC, Harris CA (1988) Nodularin, microcystin and the configuration of Adda. J Am Chem Soc 11:8557–8558.
139. Ruff TA, Lewis RJ (1994) Clinical aspects of ciguatera: an overview. Mem Queensl Mus 34(3):609–619.
140. Sasner JJ (1973) In: Martin DF, Padilla GM (eds) Marine Pharmacognosy. Academic Press, New York, pp 126–177.
141. Satake M, Murata M, Yasumoto T (1993) The structure of CTX3C, a ciguatoxin congener isolated from cultured *Gambierdiscus toxicus*. Tetrahedron Lett 32(12): 1979–1980.
142. Schantz EJ (1984) Historical perspective on paralytic shellfish poison. In: Ragelis EP (ed) Seafood Toxins. American Chemical Society, Washington, DC, pp 99–111.
143. Schneider E, Usleber E, Renz V, Terplan G (1991) Enzyme immunoassays for the detection of saxitoxin. In: Fremy JM (ed) Proceedings of Symposium on Marine Biotoxins. CNEVA, Paris, pp 145–150.
144. Shestowsky WS, Quilliam MA, Sikorska HM (1992) An idyotipic-anti-idiotypic competitive immunoassay for quantitation of okadaic acid. Toxicon 30:1441–1448.
145. Shestowsky WS, Holmes CFB, Hu T, Marr J, Wright JLC, Chin J, Sikorska HM (1993) An anti-okadaic acid anti-idiotypic antibody bearing an internal image of okadaic acid inhibit protein phosphatase PP1 and PP2A catalytic activity. Biochem Biophys Res Commun 192:302–310.
146. Shimizu Y (1978) Dinoflagellate toxins. In: Scheuer PJ (ed) Marine Natural Products. Academic Press, New York, pp 1–42.
147. Shumway SE (1990) A review of the effects of algal blooms on shellfish and aquaculture. J World Aquacult Soc 21:65–104.
148. Shumway SE, van Egmond HP, Hurst JW, Bean LL (1995) Management of shellfish resources. In: Hallagraeff GM, Anderson DM, Cembella AD (eds) Manual on Harmful Marine Microalgae. UNESCO, Paris, France, pp 433–462.

149. Sivonen K, Namikoshi M, Evans WR, Gromov B, Carmichael WW, Rinehart KL (1992) Isolation and structures of five microcystins from a Russian *Microsystis aeruginosa* strain CALU 972. Toxicon 30(11):1481–1485.
150. Soudan F (1985) Intoxications dues au phytoplancton transmises par les produits de la mer. Med Nutr 21(1):37–43.
151. Sournia A, Chretiennot-Dinet MJ, Ricard M (1991) Marine phytoplankton: how many species in the world ocean? J Plankton Res 13:1093–1099.
152. Steidinger KA (1983) A re-evaluation of toxic dinoflagellate biology and ecology. In: Round FE, Chapman DJ (eds) Progress in Phycological Research. Elsevier, Amsterdam, pp 147–188.
153. Steidinger KA (1993) Some taxonomic and biologic aspects of toxic dinoflagellates. In: Falconer IR (ed) Algal Toxins in Seafood and Drinking Water. Academic Press, London, pp 187–209.
154. Stevens DK, Krieger RI (1988) Analysis of anatoxin-a by GC/ECD. J Anal Toxicol 12:126–131.
155. Tester PA, Fowler PK (1990) Brevetoxin contamination of *Mercenaria mercenaria* and *Crassostrea virginicua*: a management issue. In: Graneli E, Sundstrom B, Edler L, Anderson DM (eds) Toxic Marine Phytoplankton. Elsevier, New York, pp 499–503.
156. Thibault P, Pleasance S, Laycock MV (1991) Analysis of paralytic shellfish poisons by capillary electrophoresis. J Chromatogr 542:483–501.
157. Tindall DR, Dickey RW, Carlson RD, Morey-Gaines G (1984) Ciguatoxigenic dinoflagellates from the Caribbean Sea. In: Ragelis EP (ed) Seafood Toxins. American Chemical Society, Washington, DC, pp 225–240.
158. Todd ECD (1990) Amnesic shellfish poisoning—a new seafood toxin syndrome. In: Graneli E, Sundstrom B, Edler L, Anderson DM (eds) Toxic Marine Phytoplankton. Elsevier, Amsterdam, pp 504–508.
159. Todd ECD (1993) Domoic acid and amnesic shellfish poisoning—a review. J Food Prot 56(1):69–83.
160. Todd ECD, Kuiper-Goodman T, Watson-Wrigth W, Gilgan MW, Stephen S, Marr J, Pleasance S, Quilliam MA, Klix H, Luu HA, Holmes CFB (1993) Recent illnesses from seafood toxins in Canada: paralytic, amnesic, and diarrheic shellfish poisoning. In: Smayda TJ, Shimizu T (eds) Toxic Phytoplankton Blooms in the Sea. Elsevier, New York.
161. Trainer VL, Baden DG (1991) An enzyme immunoassay for the detection of Florida red tide brevetoxins. Toxicon 29:1387–1394.
162. Truman P, Lake RJ (1996) Comparison of mouse bioassay and sodium channel cytotoxicity assay for detecting paralytic shellfish toxins in shellfish. J AOAC Int 79(5):1130–1133.
163. Trusewich B, Sim J, Busby P, Hughes C (1996) Management of marine biotoxins in New Zealand. In: Yasumoto T, Oshima Y, Fukuyo Y (eds) Harmful and Toxic Algal Blooms. IOC, UNESCO, Japan, pp 27–30.
164. Tubaro A, Florio C, Luxich E, Vertua R, Della Loggia R, Yasumoto T (1996) Suitability of the MTT-based cytotoxicity assay to detect okadaic acid contamination in mussels. Toxicon 34(9):965–974.
165. Usagawa T, Nishimura M, Itoh Y, Uda T, Yasumoto T (1989) Preparation of monoclonal antibodies against okadaic acid prepared from the sponge *Halichondria okadai*. Toxicon 27:1323–1330.
166. van Dolah FM, Finley EL, Haynes BL, Doucette GJ, Moeller PD, Ramsdell JS

(1994) Development of a rapid and sensitive high throughput assays for marine phycotoxins. Nat Toxins 2:189–196.
167. van Egmond HP (1989) Current situation on regulations for mycotoxins: overview of tolerances and status of standard methods of sampling and analysis. Food Addit Contam 6:139–188.
168. van Egmond HP, van den Trop HJ, Speijers GJA (1991) Worldwide regulations for marine phycotoxins. In: Fremy JM (ed) Proceedings of Symposium on Marine Biotoxins. CNEVA, Paris, pp 167–172.
169. van Egmond HP, Speijers GJA, van den Trop HJ (1992) Current situation on worldwide regulations for marine phycotoxins. J Nat Toxins 1(1):67–85.
170. van Egmond HP, Aune T, Lassus P, Speijers GJA, Waldock M (1993) Paralytic and diarrhoeic shellfish poisons: occurrence in Europe, toxicity, analysis and regulation. J Nat Toxins 2:41–83.
171. Watson-Wright W, Jellett J, Dorey M (1993) Working together. World Aquacult 24(4):26–30.
172. World Health Organization (1984) Environmental Health Criteria 37. Aquatic (Marine and Freshwater) Biotoxins. WHO, Geneva.
173. Wright JLC, Boyd RK, De Freitas ASW, Falk M, Foxall RA, Jamieson WD, Laylock MV, McCulloch AW, Mcinnes AG, Odense P, Pathak V, Quilliam MA, Ragan MA, Sim PG, Thibault P, Walter JA, Gilgan M, Richard DJA, Dewar D (1989) Identification of domoic acid, a neuroexcitatory amino acid, in toxic mussels from Eastern Prince Edward Island. Can J Chem 67:481–490.
174. Wright JLC, Bird CJ, De Freitas ASW, Hampson D, McDonald J, Quilliam MA (1990) Chemistry, biology, and toxicology of domoic acid and its isomers. In: Hynie I, Todd ECD (eds) Proceedings of a symposium on domoic acid toxicity, Ottawa, Ontario, pp 21–26.
175. Wright JLC, Falk M, Mcinnes AG, Walter JA (1990) Identification of isodomoic acid D and two new geometrical isomers of domoic acid in toxic mussels. Can J Chem 68:22–25.
176. Wright JLC, Quilliam MA (1995) Methods for domoic acid, the amnesic shellfish poisons. In: Hallegraeff GM, Anderson DM, Cembella AD (eds) Manual on Harmful Marine Microalgae. UNESCO, Paris, France, pp 113–134.
177. Yasumoto T, Murata M, Oshima Y, Matsumoto GK, Clardy J (1984) Diarrhetic shellfish poisoning. In: Ragelis EP (ed) Seafood Toxins. American Chemical Society, Washington, DC, pp 207–214.
178. Yasumoto T, Murata M, Lee JS, Torigoe K (1989) Polyether toxins produced by dinoflagellates. In: Natori S, Hashimoto K, Ueno Y (eds) Mycotoxins and Phycotoxins. Elsevier, Amsterdam, pp 375–382.
179. Yasumoto T, Murata M (1993) Marine toxins. Chem Rev 93:1897–1909.
180. Yasumoto T, Satake M, Fukui M, Nagai H, Murata M, Legrand AM (1993) A turning point in ciguatera study. In: Smayda TJ, Shimizu Y (eds) Toxic Phytoplankton Blooms in the Sea. Elsevier, New York, pp 455–461.
181. Yentsch CM (1984) Paralytic shellfish poisoning: an emerging perspective. In: Ragelis EP (ed) Seafood Toxins. American Chemical Society, Washington, DC, pp 9–23.
182. Yentsch CM (1987) Monitoring algal blooms, the use of satellites and other remote sensing devices. Aquanor 87, Trondheim, Norway.

Manuscript received July 17, 1998; accepted July 29, 1998.

Index